Alejandra Lizbeth Llontop Balarezo
Sharon Melissa Vilcarromero Ramírez

Protection and welfare of domestic animals

Alejandra Lizbeth Llontop Balarezo
Sharon Melissa Vilcarromero Ramírez

Protection and welfare of domestic animals

regarding the policies implemented by the Provincial Municipality of San Martin, Tarapoto, 2020

ScienciaScripts

Imprint
Any brand names and product names mentioned in this book are subject to trademark, brand or patent protection and are trademarks or registered trademarks of their respective holders. The use of brand names, product names, common names, trade names, product descriptions etc. even without a particular marking in this work is in no way to be construed to mean that such names may be regarded as unrestricted in respect of trademark and brand protection legislation and could thus be used by anyone.

Cover image: www.ingimage.com

This book is a translation from the original published under ISBN 978-613-9-00003-6.

Publisher:
Sciencia Scripts
is a trademark of
Dodo Books Indian Ocean Ltd. and OmniScriptum S.R.L publishing group

120 High Road, East Finchley, London, N2 9ED, United Kingdom
Str. Armeneasca 28/1, office 1, Chisinau MD-2012, Republic of Moldova, Europe
Printed at: see last page
ISBN: 978-620-7-94898-7

Contents

Dedication

This thesis is dedicated to our parents for always giving us their unconditional love, patience, dedication and motivation to achieve all our goals.

Acknowledgements
We would like to express our gratitude to our parents, friends, teachers and relatives who have contributed to this research, since their participation, experience and teachings have made it possible for us to reach this point.

SUMMARY

Protection and welfare of domestic animals with regard to the policies implemented by the

implemented by the Provincial Municipality of San Martin, Tarapoto, 2020.

Animals have always been part of our lives; even more so those that we have adopted into our homes, which is why it is necessary to legally regulate our coexistence with domestic animals at all levels of government. For this reason, we decided to investigate the relationship between the protection and welfare of domestic animals with respect to the regulations issued by the Provincial Municipality of San Martin in the district of Tarapoto, having as objectives: to identify the articles of the Law N° 30407, that prescribe on domestic animals and the knowledge of the population, to identify the policies implemented by the Provincial Municipality of San Martin, with respect to the domestic animals and the knowledge of the population of these, to identify the fulfillment of the policies implemented by the Provincial Municipality of San Martin, with respect to the domestic animals and finally to establish the relation between the protection and well-being of the domestic animals and the policies implemented by the Provincial Municipality of San Martin. Being citizens of the district of Tarapoto, our investigation was carried out in the mentioned district, with a period of execution of seven months, applying a simple correlational design, using to arrive at the results, surveys, interviews and documentary analysis guides; We obtained as a result that there are 26 articles in the Law N° 30407 that prescribe on domestic animals, being that, the majority of inhabitants know the law, with respect to the regulations implemented by the Municipality, there is only one ordinance that the majority of inhabitants do not know, manifesting and verifying that this is not fulfilled, concluding that, the relationship between the protection and welfare of domestic animals with respect to the policies implemented by the Provincial Municipality of San Martin in Tarapoto, is negative.

Keywords: Abuse, companion animal, domesticated species, public policy.

CHAPTER I

INTRODUCTION TO RESEARCH

1.1. General framework of the problem

In Peru, although there are a number of non-profit associations that safeguard animals, they do not have the support of national, regional and local authorities, and few people take an interest in and reflect on the problem of violence towards domestic animals, because there is no real awareness of the value of animal life. From our perspective, violence towards animals is an indication that it can also be directed towards human beings, which is why it is necessary to generate empathy in people towards the pain that an animal can experience, so that they become aware of the acts and omissions that threaten these living beings and thus, instead of harming, they seek to contribute to the existence of an ethical treatment of animals.

Castañeda and Hidalgo (2011), refer that when we talk about the defence of animals, we base ourselves on tolerance and respect for all forms of life. With regard to animal abuse, it is necessary to mention that, in our opinion, the conduct is more reprehensible, due to the fact that domestic animals are beings that cannot defend themselves or express themselves with words, which is why it is necessary to empathise with them and in this way be able to protect them.

Ortega et al., (2012), emphasise that animals have always been part of the human ecosystem, especially those we have adopted into our homes, and along these lines, the relationship between animals and humans has been analysed in great depth. Initially, animals were not considered within the regulations that governed human coexistence, however, this conception has evolved, to the point that some authors have come to consider them as subjects of law. Currently, there are several organisations that promote animal care and protection through awareness and indoctrination of man and society.

In this way, López (2019) indicates that at the national level, the lack of animal protection is worrying, and that for the first time on 22 May 2000 with Law No. 27265, an attempt was made to regulate the protection of wild and domestic animals kept in captivity; however, this attempt failed because it was never regulated due to technical and legal inconsistencies. Thus, after 15 years, on 16 December 2015, Law No. 30407, the Animal Protection and Welfare Law, was enacted.

On the other hand, Tarazona (2020) indicates that cruelty to animals in Peru is another form of violence that plagues society. In the quarantine, cases of animal abuse were reported on social networks, showing greater awareness and sensitisation on the part of people, as stated by Ángela Sánchez, president of the Peruvian Society for the Protection of Animals (ASPPA). In 2015, the Animal Protection and Welfare Law No. 30407 was passed, which establishes sanctions against cruel acts and mistreatment that are caused by humans. Likewise, the Peruvian National Police in 2018 created the Department of Animal Protection and Welfare in order to fight against domestic violence against pets. However, the president of Asppa confirmed that there is no particular institution in Peru that compiles statistics on the increase of animal cruelty or mistreatment.

Cervantes and Jimenes (2021), state that animal abuse is made up of a group of actions carried out of their own free will that cause physical suffering, defencelessness and even the death of an animal, whether wild or domestic.

From the above we can say that the lack of concern for regulating the protection and welfare of animals is due to the fact that the protection granted to animals in the legal sphere is currently not completely defined, because although it is true that in recent years people have become more aware of the respect for animals, showing great progress on this issue, It is also true that the cases of abandonment and cruelty to animals has not yet decreased, this can be seen day by day, walking down a street and seeing animals that despite being in a pitiful state are not helped, also, in social networks it is not new that daily cases of abandonment and abuse of both wild and domestic animals are denounced.

In the city of Tarapoto it is enough to go out into the streets to notice the indifference and cruelty with which stray domestic animals are treated. As we know, one of the most common forms of transport in our city is the motorbike, so unfortunately it is very common to see domestic animals being run over on the streets on a daily basis, without there being a culture of denouncing these incidents, let alone helping them.

This is why Cruz (2019) emphasises that caring responsibly for a pet means that the owner is responsible for providing the necessary conditions of a suitable environment adapted to its needs, which should include food, housing, shelter, medical care, as well as appropriate affection.

Pets share a special bond with humans, contributing both positively and negatively to the environment, as well as to the physical and mental health of a person, which is why pets often end up being considered as an integral part of the family nucleus due to the innocent love they show, thus accustoming them to the world of humans.

1.2. Formulation of the research problem

The problem on which our research is based is as follows:

What is the relationship between the protection and welfare of domestic animals and the policies implemented by the Provincial Municipality of San Martin, Tarapoto, 2020?

1.3. Research hypothesis

The relationship between the protection and welfare of domestic animals and the policies implemented by the Provincial Municipality of San Martin in the district of Tarapoto, 2020 is positive.

1.4. Objectives

1.4.1. General objective

To determine the relationship between the protection and welfare of domestic animals with respect to the policies implemented by the Provincial Municipality of San Martin, Tarapoto, 2020.

1.4.2. Specific objectives

A.   Identify the articles of Law N° 30407 that prescribe on domestic animals and the knowledge of the population of the district of Tarapoto about them.

B.   To identify the policies implemented by the Provincial Municipality of San Martin in the district of Tarapoto, with respect to domestic animals and the knowledge of the population of the district of Tarapoto about them.

C.   To identify compliance with the policies implemented by the Provincial Municipality of San Martin in the district of Tarapoto with regard to domestic animals.

D.   To establish the relationship between the protection and welfare of domestic animals and the policies implemented by the Provincial Municipality of San Martin in the district of Tarapoto, 2020.

1.5. Rationale for the research

Theoretical justification

With this thesis report we seek to demonstrate the existence of a problem in relation to animal abuse, in our theory, generated by the ignorance and lack of interest of the population of Tarapoto, as well as by the competent authorities. For this reason, the need to investigate the relationship between the protection and welfare of domestic animals with respect to the policies implemented by the Provincial Municipality of San Martin in Tarapoto was born.

Methodological justification

This research aims to achieve the objectives of the survey instruments, documentary analysis guide and interview guide, as these will allow us to identify the interest and knowledge of the population of Tarapoto regarding Law No. 30407, the Animal Protection and Welfare Law, which is fundamental in order to curb mistreatment and cruelty to domestic animals, and is based on the criteria established at the National University of San Martín - Faculty of Law and Political Science.

Practical justification

The following investigation finds practical support in the need to stop animal abuse and to contribute favourably to the implementation of measures to benefit animals, which is why it is necessary to investigate the existence of this problem, as it is an issue that covers both animal welfare and public health, which is why as a society we must take an interest and apply the mechanisms for the protection of animals, taking into account the Law N°30407 - Law for the Protection and Welfare of Animals.

CHAPTER II

THEORETICAL FRAMEWORK

2.1. Background to the research

Burgos (2019), in his degree thesis entitled "Guidelines for the public policy of animal welfare in the Municipality of Sibaté (Cundinamarca) Colombia" by the Pontificia Universidad Javeriana, which sought to generate guidelines for the implementation of an effective public policy favouring companion animals in the Municipality of Sibaté, Colombia, specifically dogs and cats, using NATO methodology (Hood 1986 cited in Fontaine), Colombia, specifically dogs and cats, using the NATO methodology (Hood 1986 cited in Fontaine), which takes into account instruments of information (nodalidad), authority (authority), treasure (treasure) and organisation (organisation) in order to obtain a comprehensive design of public policy. The result is a solid public policy proposal that favours the protection of animals. Concluding that:

At all levels of government, animals should be part of public policies, as it affects a public health issue, being that the Municipality of Sibaté, should relate the administrative part of the Municipality with animal protection organisations, society, private companies when they formulate and implement their policies, ensuring animal welfare. It should improve the dialogue with the various actors, with the aim of creating actions that lead to the formulation and implementation of public policies.

Butriago (2019) is a graduate thesis entitled "Promotion and communication in animal welfare and responsible pet ownership in Colombia, by the University CES- Colombia", whose objective was to:

To identify the actions referred to the promotion and communication about animal welfare and responsible pet ownership in Colombia, in order to argue for a reform of strategies that can be added to public policies in the country mentioned. The population and the sample are based on the publications and experiences of the Ministry of Health, Environmental Health Secretariats at municipal and departmental level, foundations dedicated to animal protection and veterinaries, with documentary research being the type of study applied. The conclusion is that animal welfare and the environment are in danger due to a lack of knowledge, awareness and sensitisation.

Vera (2020), in her degree thesis entitled "Perception of the owners on the current laws of protection and animal ownership, applied to dogs treated in two veterinary practices in the city of Guayaquil", had as her objective:

To evaluate the way in which the owners of dogs in Guayaquil perceived the laws in force regarding animal protection and ownership, carrying out surveys of 102 owners, using descriptive statistical analysis, with the result that 67% of the people surveyed had a medium level of knowledge of the regulations in force regarding animal protection and ownership, concluding that it is necessary to carry out campaigns aimed at socialising the laws that already exist.

Moron (2019), in his research "La controversia del delito de Abandono y actos de crueldad contra animales domésticos y silvestres tipificado en el art. 206-A del código penal con la Ley N° 30407 respecto a su calificación como seres sensibles (tesis de pre-grado) por la Universidad Andina del Cusco", whose objective was to determine the controversy between the crime of Abandono y Actos de crueldad contra Animales Domésticos y Silvestres, established in art. 206-A del Código Penal, with the law 30407 regarding sentient beings. This research had a qualitative approach, since it took into account the analysis and argumentation of the documentary information collected for the elaboration of conclusions. Thus concluding that:

Due to the existing controversy between Law N° 30407 and article 206-A of the Code, due to the classification of animals as sentient beings, it is necessary to modify the civil, penal, administrative or institutional norms, all those referring to animal protection and welfare, as well as animal mistreatment and cruelty, since they contradict each other in the qualification that they give to animals, confusing the protection of animals with the economic purpose that the regulations seek in our country. This is the only way to avoid the ineffectiveness of the regulations and to avoid putting at risk the legal insecurity that currently exists in relation to the protection of animals.

Luke (2019), in his research "Analysis of the situation of companion animals in a state of abandonment and the implementation of a municipal legal regime of protection in the district of Paucarpata (pre-graduate thesis) by the Universidad Autónoma San Francisco", seeking to investigate the situation of companion animals in a state of abandonment and implementation of a Municipal Legal Regime in the district of Paucarpata, which guarantees the safety, of the population and of the abandoned animals, employing information gathering techniques, using document analysis guides and also the review of current regulations; thus concluding that:

In order to protect domestic animals, the central government enacted legal mechanisms such as Law N°30407, however, in spite of this, both in Arequipa and its districts, due to the lack of regulations implemented by their local governments, they have not been able to prevent the overpopulation of canines, the most affected being Paucarpata, which is a scene of attacks by packs of dogs on society, contamination by animal faeces and unhealthy effects, causing people to abandon their animals.

Chú (2016), in his undergraduate thesis entitled "Diagnosis on responsible pet ownership in thirteen sectors of the district of Tarapoto- Province of San Martin, by the National University of San Martin", aimed to carry out a diagnosis on Responsible Pet Ownership (TRAC) in the homes of the thirteen sectors of the district of Tarapoto applying a non-experimental design of sectional type, performing a descriptive analysis, obtaining as a result:

Regarding animal mistreatment, that in the Punta del Este Sector this problem is palpably observed, due to the fact that the respondent had a pet living in the street, without even knowing about the schedule of deworming and less about the care that a pet should have, finally recommends informing ourselves before adopting a pet, enforcing the law No. 30407 and encouraging the participation of both public and private entities.

Cavalcanti (2016), in his research "Abandonment of pets, followed by death and criminal liability under Law No. 30407 in Peru (Undergraduate Thesis) by Alas Peruanas University. Tarapoto", aimed to establish whether the abandonment of pets followed by death has an impact on the criminal liability of the owner according to Law No. 30407. The research was basic and descriptive, with a population of criminal lawyers and using interviews and textual files as data collection techniques and instruments. Thus concluding that:

The Peruvian State enacted Law N°30407 for the benefit of domestic animals, so that the legal system makes the owner criminally responsible for abandonment, leading to the death of domestic animals, but, in spite of this, animals continue to be victims of violence, exploitation and extermination; likewise, the legal system is incompetent since society, as well as civil servants and public officials, do not know most of their rights, in spite of the existence of previously enacted laws.

2.2. Theoretical underpinnings

2.2.1. Protection and welfare of domestic animals

2.2.1.1.　　Concept

a.　Animal welfare

According to the Animal Protection and Welfare Act (2015), this includes the quality of animal life, respect for the species and natural habitat, and man must adapt the spaces provided to them, so that these animals can develop while maintaining their original behaviour, in order to safeguard their physical and mental health. Likewise, in 1997, for the authors Duncan and Fraser, when talking about animal welfare, the following should be considered: The psychological domain, such as pleasure and suffering, the biological aspect, such as the state of health, the nature of the species and ensuring that they develop their behaviour in its full range.

On the other hand, Fiedrich (2012) states that it is necessary to consider the "5 freedoms" for animal welfare, formulated in 1993 by the Farm Animal Welfare Council of the United Kingdom (Farm Animal Welfare Council):

1.　Freedom from hunger and thirst: "This is achieved by providing access to clean water and a diet that allows them to enjoy good health.

2.　Free from discomfort: "Implies that animals should be provided with an environment appropriate to their needs, including protection and comfortable resting areas".

3.　Freedom from pain, injury and disease: "This is achieved by structuring on-farm preventive schemes, as well as timely diagnosis and treatment.

4.　Free to express their normal behaviour: "We must provide sufficient space, adequate infrastructure and companionship with animals of the same species, so that they can live together".

5.   Freedom from fear and stress: "We must ensure conditions that avoid psychological suffering of animals".

b.   Human Health: The World Health Organisation (1946) defines health as "a state of complete physical, mental and social functioning and not merely the absence of disease or infirmity".

Likewise, the Congress of the Republic (1997) indicated through the General Health Law N° 26842 (1997) that: IV. Public health is the main responsibility of the State. That responsibility is shared between: Individual, society and the State. IX. Health regulations are of public order and regulate health, as well as protect the environment for well-being and medical care, with the aim of enabling people to recover and rehabilitate themselves.

c.   Human and animal health: Human health, whether for physical, cultural or psychological reasons, depends on a variety of aspects of animal health, such as those animals used for production, livestock or poultry, as well as pets that have a physiological and psychological impact on the health of their owners. "What would humans be without animals? If there were no animals, man would perish by affliction of spirit, because what happens to animals in the future will happen to man; there is a connection in everything. That question was asked by Chief Seattle in 1864 in the letter he wrote to the President of the United States.

2.2.1.2.   Theory on the protection and welfare of domestic animals

Theory of the moral status of animals by Peter Singer.

His theory is based on "sentience", noting that this capacity refers to the fact that both humans and animals are the same.

Animal Law Theory by Tom Regan.

His theory was based on the following, that humans have rights just like many animals, so it is said that humans who torment animals also torment other humans.

2.2.2. Public policies

Within public policies, two points were taken into account: the first is Law No. 30407 - Animal Protection and Welfare Law, as this is the law from which all locally applied regulations on animal protection and welfare derive, and the second is Municipal Policies in accordance with the aforementioned law.

2.2.2.1.   Law No. 30407 on Domestic Animals: Law on the Protection and Welfare of Animals, 2015

Law No. 30407 (2015) is a regulation issued by the legislature on 16 December 2015. This law conceptualises the meaning of animal protection and welfare as a conglomerate of fundamentals, which allude to the quality of life that can be offered to animals, whether with respect to the protection of species, as well as the adaptation of these to the environment that man has given them for their development, so that their behaviour can be expressed naturally, thus manifesting a complete state of health, both physical and mental.

Articles of Law No. 30407 prescribing on the protection and welfare of domestic animals.

Principles: The principles of: the principle of animal protection and welfare, the principle of biodiversity protection, the principle of integral collaboration and responsibility of society, the principle of harmonisation with international law, and the precautionary principle are established. (Art. 1 of Law No. 30407).

Purpose: Its purpose is to ensure the comfort and protection of all wild and domesticated vertebrate animals in captivity, within the framework of measures to protect life, animal health and public health. (Art. 2 of Law No. 30407).

Object of the law: The object of the aforementioned law is to safeguard the life and health of vertebrate animals, whether domestic or wild; it also seeks to curb all actions involving mistreatment and cruelty that cause unnecessary pain, injury or death; in addition to this, it also aims to promote through education, respect for life and animal welfare. (Art. 3 of Law No. 30407).

Duties of individuals: Everyone is responsible for the protection and comfort of the animal; an adult is responsible for the purchase and keeping of the animal; and the responsible pet owner must be provided with adequate habitat, adequate food and specialised medical care (Art. 5 of Law No. 30407).

Duties of the State: It is the duty of the Peruvian State to establish the necessary mechanisms to protect domestic animals, to provide guarantees for their health, life and harmonious coexistence with their environment; it must also establish suitable treatment and adequate zootechnical management of farm animals and, finally, it must promote the defence and sustainability of wildlife, in accordance with Law No. 30407 (Art. 7 of Law No. 30407).

Governing body and inter-sectoral coordination: The Ministry of Agriculture and Irrigation will act as the governing body issuing complementary regulations to regulate the protection and welfare of domestic and wild animals in captivity and when used for experimentation, research, education, conservation and commercialisation; they design and develop guidelines in conjunction with the Ministry of the Environment. The Ministry of Agriculture and Irrigation manages the contribution of the different sectors to the protection and welfare of animals (Art. 9 of Law N° 30407).

Animal welfare and protection committees: Local and regional governments should establish a committee in charge of animal welfare and protection. This committee must be made up of the governor of the region and the mayors of the provinces, as well as a representative of the professional associations of doctors, biologists and veterinarians of our country. (Art. 11 of Law N° 30407).

Animals as sentient beings: This law recognises domestic and wild vertebrate animals kept in captivity as sentient beings (Art. 14 of Law 30407).

General prohibitions: A. Leaving an animal on the road is a cruel act and also becomes a danger to public health; B. The use of animals for public or private recreation indicates that the animal is forced or subjected to activities that are incompatible with its nature or affect its physical integrity and well-being; C. The keeping, hunting, capture, breeding, purchase and sale for human consumption of animal species that are not considered farm animals, except for wild species that are kept for the purpose of production or consumption in zoos or in management areas approved by the competent bodies, as well as those that are obtained through hunting for the purpose of survival by aboriginal communities; D. Fighting of domestic animals and wild animals in public or private places. (Art. 22 of Law No. 30407).

Measures for the protection and welfare of companion animals or pets: This article warns about the measures for the protection and welfare of pets, organised by the Ministry of Health in harmonisation with the Ministry of the Environment, establishing that all persons and entities that are owners of domestic and wild animals must carry out these actions for the protection and welfare of animals, which are based on good practices (Art. 21 of Law No. 30407).

Prohibition of violence against pets: This section establishes prohibitions of violence against animals, where all practices that could oppose the protection and welfare of domestic animals are contemplated (Art. 27 of Law No. 30407).

2.2.2.2.   *Municipal policies*

According to Law No. 27972 (2003), the Law on Municipal Policies defines municipal policies as those whose objective is to ensure justice and peace in society, in the municipal territory, the political representation of the inhabitants, applying municipal regulations and promoting the intercession of conflicts, through dialogue and social prevention, as well as directing through these policies the socio-economic, integral, sustainable and harmonious progress of its jurisdiction.

According to Decree Law N° 26162 (1992), Law of the National Control System, in Peru, municipal policies refer to the "group of actions by which the entities are going to achieve their aims, objectives and goals, which are included in the governmental policies imposed by the Executive and Legislative Power".

Municipal regulation "Municipal Ordinance N°014-2020-MPSM".

The Municipal Ordinance N°014-2020-MPSM authorises the regulation on the regime of keeping and registration of dogs in the district of Tarapoto, which contains 36 articles in reference to the object and scope of application, establishing the Legal Regime for the Registration and Keeping of dogs, The regulation establishes the legal regime for the registration and ownership of dogs, the use of public areas by dogs in the district of Tarapoto, with the aim of protecting the physical integrity of people and the welfare of domesticated animals, so that they can develop in an environment that provides them with the necessary living conditions, imposing sanctions on anyone who violates these regulations.

Case law

According to the Constitutional Court System (2018), it points out as jurisprudence to take into account the Case N°1413-2017- Lima, from which the TC points out that pets that act as a guide, since in Antonio Miro Quezada which is a building pets were prohibited, so, the issuance of the ruling resolution is based centrally on recital number 19 which, refers on the failure to overcome the proportionality test, contemplating articles 35.8.1 and 15.8.3. The restriction of having pets in the building, of taking in new pets and using the lift together with them, is not proportional and affects the rights to free transit and the free development of personality. In addition, recital number 20 on the deprivation of the entry or stay of visitors with animals in the building points out that article 35.8.2. of the building regulations must consider the existence of animals such as guide dogs, which cannot be prevented from entering the building where the plaintiff lives, as well as the private areas; based on the aforementioned recitals is that the collegiate ordered to modify the internal regulations of the aforementioned building, since domestic guide dogs help people with different abilities, as stated in Law No. 30407.

On the other hand, with regard to animal cruelty, we have file N° 06261-2020, in which the Fifth Supraprovincial Unipersonal Criminal Court of Chiclayo and Ferreñafe (2021), states that from the evidence presented by the complainant (pet owner), based on its recital number 5.1, which states that, having proved the facts of the trial against the accused, the Fifth Criminal Court confirms that the qualification made by the defence of the injured party is correct and the facts under discussion are within the crime against property, in the modality of acts of cruelty against domestic animals, foreseen and punished by article 206-A° first paragraph of the Criminal Code; also, based on recital number 6.2. which states that the acts carried out by the accused are accredited and that said conduct is immersed within the aforementioned offence; also referring to recital number 8.1. which indicates the inexistence of grounds to justify the act in question in the trial, as well as recital number 8.2. which indicates that the facts were perpetrated by the accused in full use of her reason, mentally stable and with the capacity to act in a different way to the one carried out, being so, the Criminal Court condemned the accused for the crime of acts of cruelty against domestic animals.

CHAPTER III

MATERIALS AND METHODS

3.1. Scope and conditions of the research

3.1.1. Context of the research

This research took place in the district of Tarapoto, province and department of San Martin, located at 333 m.a.s.l., in the Peruvian Amazon; it has an area of 4 500 hectares and an approximate population of 73 015 inhabitants (MPSM, 2022).

This district was chosen for the research because it is the district where the authors live, and because it is the most populated and developed district in the San Martin region. At present, almost no research has been carried out in the city regarding the protection and welfare of domestic animals, so, as it is a district with great potential for development and a visible lack of interest in stray domestic animals, it was essential for the thesis students to carry out the study. Likewise, the regulations issued by the Provincial Municipality of San Martín (MPSM) regarding domestic animals will be studied, as the population of the district of Tarapoto is governed by the ordinances issued by this entity.

3.1.2. Implementation period

The research was carried out over a period of 7 months; the activities were executed as follows:

Table 1

Timeline for specific objective 1.

Specific Objective N°1: To identify the articles of Law N°30407 that prescribe on domestic animals and the knowledge of the population of the district of Tarapoto about them.

Activity	Month 1	Month 2	Month 3	Month 4	Month 4	Month 5	Month 6	Month 7
Analysis of the Law N°30407.	X							
Selection of the articles of Law No. 30407 that prescribe on domestic animals and analysis of them through the Documentary Analysis Guide.					X			
Conduct 200 surveys and 100 surveys in person at the Plaza				X	X			
The aim of this study was to identify the population's knowledge of Law N° 30407 and the articles it contains.								

Table 2

Timeline for specific objective 2.

Specific Objective N°2: To identify the policies implemented by the Provincial Municipality of San Martín in the district of Tarapoto with regard to domestic animals and the knowledge of the population of the district of Tarapoto about them.

Activity	Month 1	Month 2	Month 3	Month 4	Month 5	Month 6	Month 7
Gather information about the regulations issued by the MPSM regarding the protection and welfare of domestic animals, through the institutional portal of the municipality and by request to the entity.	X X						
Analyse the Municipal Ordinances that prescribes on							

protection and welfare of domestic animals through the documentary analysis guide.

Table 3

Timeline for specific objective 3.

Specific Objective N°3: To identify compliance with the policies implemented by the Provincial Municipality of San Martín in the district of Tarapoto with regard to domestic animals.

Activity	Month 1	Month 2	Month 3	Month 4	Month 5	Month 6	Month 7
Interviews with specialists in the field.	X						
Conduct 200 surveys and 100 surveys							
The project was implemented in the Plaza de Tarapoto and the Suchiche park in the district, consulting on compliance.	X	X					

Table 4

Timeline for specific objective 4.

Specific Objective N°4: To establish the relationship between the protection and welfare of domestic animals and the policies implemented by the Provincial Municipality of San Martín in the district of Tarapoto, 2020.

Activity	Month 1	Month 2	Month 3	Month 4	Month 4	Month 5	Month 6	Month 7
Analysis of statistical results.					X			
Interview analysis.					X			
Establishing relationships.							X	
Issuance of the final report.								X

3.1.3. Authorisations and permits

In the present thesis, no means of investigation subject to access restriction or control of standards are involved, which is why this does not apply.

3.1.4. Environmental monitoring and safety protocols

Not applicable in this thesis.

3.1.5. Application of international ethical principles

I. Integrity. This research was carried out with the aim of contributing with new knowledge The aim of the project is to raise awareness among the population of the district of Tarapoto and to highlight a problem that often, without realising it, ends up involving us. For this reason, the thesis students undertook the commitment to provide true and valuable information that contributes to the eradication of animal abuse.

II. Respect for people. This principle was present throughout the development of the research, because in the first place the study responds to a social need, whose purpose is to contribute in a positive way to each person who lives in the district of Tarapoto. At the time of applying each interview and/or survey, it was stated that participation was voluntary and contributed to a research work with academic purposes; likewise, each opinion and answer given was significantly valued, respecting each point of view and analysing it in a transparent manner, with the aim of carrying out a truthful investigation.

III. Justice. This principle is present, as the opinion of not only the ordinary citizen of the city of Tarapoto was valued, but also the judgement of experts in environmental matters was added to these opinions, resulting in equitable responses, as this allows us to objectively analyse the different points of view and perceptions of the inhabitants of the district of Tarapoto.

IV. Scientific validity. We found this principle throughout the research, as information was collected from reliable research works, and we proceeded to cite each author taken as a reference, using the APA Standards sixth edition.

3.2. Variable system

According to Pérez (2007), variables are the fundamental units of the scientific method. A variable is considered to be anything that has its own characteristics and is susceptible to modifications or changes, and can also be studied, controlled or measured in a research study.

Within the research, abstract variables were proposed and formulated, as the study was carried out on immaterial aspects, which are detailed as follows:

Variable 1: Animal protection and welfare.

Table 5

Description of specific objective variable N° 01.

Specific Objective N° 01. To identify the articles of Law N° 30407 that prescribe on domestic animals and the knowledge of the population of the district of Tarapoto about them.

Abstract variable	Specific variable	Means of Registration	Unit of measurement
Protectionand animal welfare.	It cannot be measured.	Analysis of Law No. 30407 (Documentary analysis guide) and documentary analysis) and Direct procedure through the use of data collection instruments (surveys).	Ordinal

Table 6

Description of specific objective variable N° 02.

Specific Objective N° 02. To identify the policies implemented by the Provincial Municipality of San Martín in the district of Tarapoto with regard to domestic animals and the knowledge of the population of the district of Tarapoto about them.

	Abstract variable	
	Concrete variable	
Unit of measurement		
	Means of Registration	
Protection y animal welfare.	Analysis of the Municipal Ordinance N° 014-2020-MPSM and (guide of Documentary analysis is not possible) O di l direct procedure by means of the ʳ ⁿᵃ use of data collection instruments (survey).	

Variable 2: Public policies.

Table 7

Description of specific objective variable N° 03.

Specific Objective N° 03. To identify compliance with the policies implemented by the Provincial Municipality of San Martín in the district of Tarapoto with regard to domestic animals.

Abstract variable	Specific variable	Means of Registration	Unit of measurement
Policies implemented by the	It cannot be	Direct procedure	Ordinal

Provincial Municipality of San Martin.	measured.	by using information gathering instruments (survey), expert judgement (interview guide) and analysis of Law N° 30407 and Municipal Ordinance N° 014-2020-MPSM (documentary analysis guide).

Table 8

Description of specific objective variable N° 04.

Specific Objective N° 04. To establish the relationship between the protection and welfare of domestic animals and the policies implemented by the Provincial Municipality of San Martin in the district of Tarapoto, 2020.

Abstract variable	Variable concreta	Means of Registro	Unit of medida
Policies implemented by the Provincial Municipality of San Martin.	Cannot be expert ($_{Ordinal}$ analysis [r na]	Direct procedure through the use of information-gathering instruments (survey), judgement Measurement Guide. N° 30407 and Municipal Ordinance N° 014-2020- MPSM (Documentary analysis guide).	interview) and law

3.3. Research design

3.3.1. Type and level of research

The present research is quantitative because it has a schematic method for collecting and processing information acquired from various sources. According to Fernández and Días (2002), this process is carried out using statistical and mathematical tools with the aim of quantifying the research problem.

The research design is at the correlational level, the correlational level has the fundamental purpose of finding out the existing relationship between two or more concepts. Rus (2020) states that here the researcher basically measures two variables, interprets them and determines the existing statistical relationship; it can be said that he/she tries to discover how one variable varies when the other does too.

3.3.2. Population and sample

The population consisted of the citizens of the district of Tarapoto and the sample taken was a simple random sample of 300 people.

3.3.3. Analytical, sampling or experimental design

3.3.3.1. Non-experimental design

Legal research is essentially a non-experimental design because it is systematic and empirical, where variables are not manipulated because they have already occurred. Inferences about the relationships between variables are made without direct intervention or influence, and these relationships are observed as they occur in natural contexts. This is represented graphically as follows:

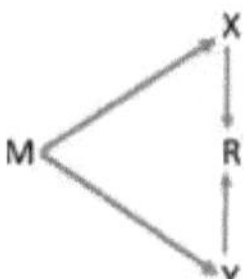

Where:
M : Research   sample
X : Protection and welfare of domestic animals
Y : Public policies

Table 9
Data collection techniques and instruments.

Variables	Techniques	Instruments	Informants
Protection and welfare of domestic animals	Survey. Documentary analysis guide.	Survey. Documentary analysis guide.	Residents of the district of Tarapoto. Provincial Municipality of San Martín.
Public policies	Guide to interview.	Guide to interview.	Non-profit association Yupi Wasi. MPSN Competent Authority

3.3.3.2.   *Statistical analysis*

For the development of the research and as it is a quantitative approach, the statistical methods of "SPSS" were used, which in its dimension is a statistical computer programme that is used in these types of research, as well as Microsoft Excel spreadsheets.

3.4. Investigation procedures

3.4.1. Specific Objective 1

To develop this objective, first the survey was prepared (taking into account all the objectives set out) and the documentary analysis guide of Law N°30407, and then proceeded to analyse the Law itself; after this, the survey was applied to the inhabitants of the district of Tarapoto; 100 surveys were applied in person and 200 were applied virtually through a Google questionnaire, both for this objective and for the others.

3.4.2. Specific Objective 2

To develop this objective; and on April 10, 2023, to corroborate the information found on the website of the MPSM, a request for access to public information was submitted to this entity, in which the Municipality was required to provide information on all regulations issued by the same regarding the Protection and Welfare of Domestic Animals, with a document dated April 13, 2023, they informed that only the Municipal Ordinance N°014-2020-MPSM existed on this subject, which prescribes the regime of ownership and registration of dogs in the district of Tarapoto, which is why we proceeded to analyse it; Based on this, we proceeded to elaborate and develop the documentary analysis guide of the ordinance and applied the survey.

3.4.3. Specific Objective 3

To develop this objective, we also used the survey applied to the population of the district of Tarapoto, we took into account the analysis of the ordinance issued by the Municipality, we elaborated the interview guide based on the survey applied to the inhabitants and we preceded to obtain the judgement of experts.

3.4.4. Specific Objective 4

In order to develop this objective, the survey, the documentary analysis guide and expert judgement were used and analysed.

20

CHAPTER IV

RESULTS AND DISCUSSION

4.1. Results

4.1.1. According to Specific Objective 1

Specific Objective N° 1: To identify the articles of Law N° 30407 that prescribe on domestic animals and the knowledge of the population of the district of Tarapoto about them.

4.1.1.1.   Results of the survey applied to the problems of the district of Tarapoto.

The questions aimed at answering this objective are questions 1 and 2, which yield the following results:

1.   Do you know if there is any law that protects and provides welfare for domestic animals?

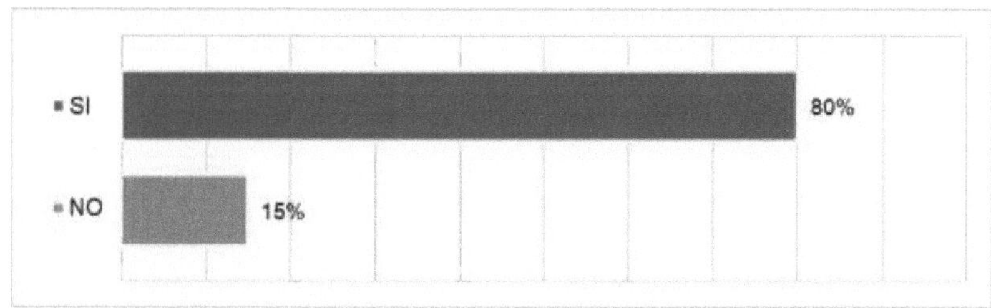

Figure 1

Do you know if there is any law that protects and provides welfare for domestic animals?

Source: Own elaboration

From Figure 1 we can see that 15% of the surveyed population did not know of the existence of any law that protects and provides welfare for domestic animals, while 85% said they knew of the existence of any law that protects and provides welfare for domestic animals.

2.   Are you aware of the existence of the Animal Protection and Welfare Law - Law N°30407 and its articles referring to domestic animals?

Figure 2

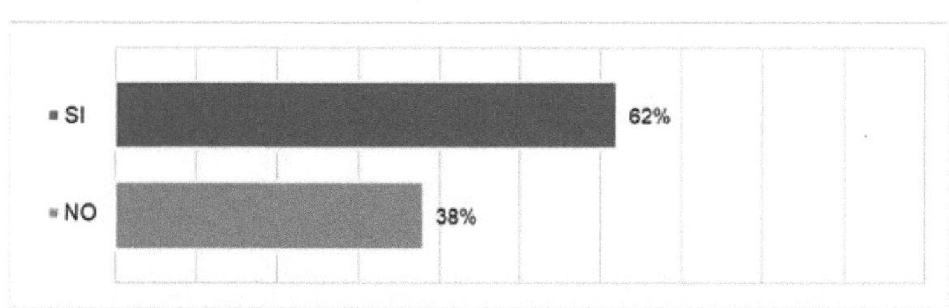

Are you aware of the existence of the Animal Protection and Welfare Law - Law *N°30407 and its articles referring to domestic animals?*

Source: Own elaboration

From Figure 2 we can see that 38% of the surveyed population does not know of the existence of Law N°30407 and its articles referring to domestic animals, while 62% indicate that they do know of the existence of the Law for the Protection and Welfare of Animals and its articles referring to domestic animals.

4.1.1.2. According to interviews with experts in the field of protection and welfare of domestic animals

The questions aimed at answering this objective are questions 1, which yield the following results:

2.   Could you comment on Law No. 30407 - Law on the Protection and Protection of the Rights of the Child?

Animal Welfare, with regard to domestic animals?

Table 10
Question 2.

Expert response	
Marco Antonio Sánchez Huaripata (Deputy Manager of Local Economic Development of the MPSM)	**Elizabeth García Panduro (Representative of the non-profit association Yupi Wasi)**
The law contemplates the conditions that a pet must have under our care, based on this, the law indicates that we must have the five freedoms as the main axis, the same that we have to work on; the first is freedom from hunger and thirst, to have adequate conditions, not to limit their functions as pets either. The law also contemplates animal comfort, the conditions that the environment must have; by that I mean that if I have a flat I cannot have a Labrador, a Doberman, etc., I can suddenly have a Chihuahua, a lap dog, whatever is best suited to our spaces. You also have to be aware of what you have and what you can give to your pet, because that is also part of animal welfare.	Well, I know the law by name, I am glad to know that at least there is a law that is trying to promote the issue of protection. Although it is true that most people do not know in detail what it is about, at least it is there, and I hope that little by little through these types of projects, through the associations, through the Municipality itself, more information can be given about this law and above all how it is applied, because the law may be there but many times we do not know how to proceed, what we have to do or what is considered an infraction or when I can make a complaint and that type of thing, but at least it is there, it is a first step.

4.1.1.3. According to the documentary analysis guide of the Law N°30407 - Law on Animal Protection and Welfare regarding the articles prescribing on domestic animals

The following result is obtained:

1.   Law N° 30407: "Animal Protection and Welfare Law", has 26 articles that are applicable to the protection and welfare of domestic animals, 6 of these articles establish the role and intervention of the local government in the application of this law.

2.   Law No. 30407, with regard to domestic animals, regulates the responsibility of persons and the State, the governing body and executing and supporting bodies, animal protection and welfare associations, ownership, protection and management, euthanasia of domestic animals and finally infringements and penalties, in case of non-compliance with the Law.

3.   On the intervention and role of local and regional governments, we have to the law states:

a) They have a duty to address complaints and intervene to enforce the law, b)	They should encourage the creation and also the operation of temporary shelters for abandoned pets.

c)	To monitor the application of the law and supervise compliance with it when granting licences and authorisations related to the keeping, marketing, transport and veterinary care of animals.

d)	They should be part of the animal protection and welfare committee.

e)	They have the power to apply the necessary mechanisms to stop animal abandonment and to impose the corresponding sanctions.

f)	Finally, it establishes the sanctioning power of local and regional governments within the scope of their material and territorial competences, as well as the coercive enforcement of the obligations arising from the law.

4.1.2. According to Specific Objective 2

Specific Objective N° 02: To identify the policies implemented by the Provincial Municipality of San Martín in the district of Tarapoto with regard to domestic animals and the knowledge of the population of the district of Tarapoto about them.

***4.1.2.1.	Results of the survey applied to the problems of the district of Tarapoto.***

The question aimed at answering this objective is question 3, which yields the following results.

3.	Do you know of any decree or ordinance that the authorities of the municipality Sint Maarten Provincial Council have issued with regard to Animal Protection and Welfare?

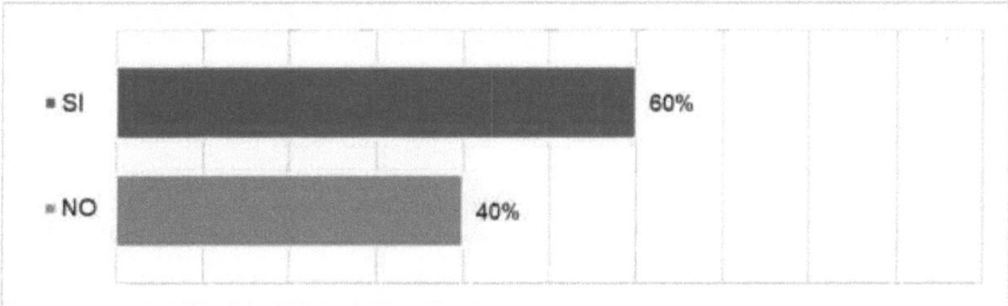

Figure 3

Do you know of any decree or ordinance that the authorities of the Provincial Municipality of San Martin have issued regarding Animal Protection and Welfare?

Source: Own elaboration

From Figure 3 we can see that 40% of the population say that the Municipality has not issued any decree or ordinance regarding animal protection and welfare, while 60% indicate that the Municipality has issued a decree or ordinance regarding animal protection and welfare.

***4.1.2.2.	Results of the interview with experts in the field of protection and welfare of domestic animals.***

The questions aimed at answering this objective are questions 3 and 4, with the following result:

3.	Has the Provincial Municipality of San Martin currently taken any action in this regard?

action with regard to Law No. 30407?

Table 11
Question 3.

Expert response	
Marco Antonio Sánchez Huaripata (Sub-Gerente de Desarrollo Económico Development of the MPSM)	**Elizabeth García Panduro (Representative of the non-profit association Yupi Wasi)**
Of course, based on this law that we have, we have already enacted in 2020, the Municipal Ordinance N° 014-2020 which, contemplates responsible pet ownership here in San Martin, regulates under what conditions we must have them, how we have to have them and the evaluations that we have to pass to have a pet; the authorization implies the registration of our canine in the municipalities; because, the law contemplates that local governments are in charge of keeping the pet registry and that also implies the type of pet that I am going to have, that is to say, our ordinance contemplates that if I am going to have pets considered potentially dangerous, I have to have special authorisations, among them, a psychological authorisation. It says a lot about how we keep our pets, now we also have some inspections due to some photographic complaints made by neighbours, we go with the inspection area based on our ordinance to regulate this a little; now we do not see many dogs in Tarapoto living on the roof, but we do have dogs that still spend the night at the door of the houses, on the public road and this is an issue that we have to regulate, obviously this is a very big public health problem.	There is an ordinance and I know that this new administration is a little more committed to this process. Although it is very slow because of the bureaucracy, I imagine, but at least I know that they are working, for example, in the control, I know that a person can go to the Municipality and report a case of abuse, abandonment, neglect or whatever a person sees and considers that it is not right with respect to a pet, they can go, I do not know exactly what area, I don't know exactly what area, but I know that there is an area that oversees this, the Municipality sends them to check to see what is happening, if they are living in a bad situation, sometimes we know that they are living on the roof without water and food, or also as we see that they are being raised in the street, I have also seen that some people let them loose to go for a walk and in some cases pets have been run over because of this; I know that a person can go and denounce and there is a process, but I don't know what else happens after that, but I do know that at least a prosecutor goes, gives certain actions to the police, and then they can take action. I also know that the Municipality is trying to work in other areas to improve in terms of responsible pet ownership and sanctioning in case of neglect, abandonment or mistreatment.

4. Is there any decree or ordinance that the authorities of the Municipality
What has the Provincial Council of Sint Maarten issued regarding Animal Protection and
Welfare? Let us know.

Table 12

Question 4.

Expert response	
Marco Antonio Sánchez Huaripata (Sub-Manager for Economic Development of the Local MPSM)	**Elizabeth García Panduro (Representative non-profit association Yupi Wasi)**
Of course, as I mentioned at the beginning, we have been working for some years now focusing on responsible pet ownership, and that is why the Ordinance 014- 2020-MPSM was born, which regulates responsible pet ownership, if we have it regulated here in the Province of San Martin, in Tarapoto we basically have an Ordinance that contemplates all of this.	I am aware of Ordinance N°14-2020, I know that not picking up faeces, not walking without a leash or without a muzzle if necessary, veterinary care, proper feeding can be sanctioned; but I don't know how much people know about this, beyond the number which is important to know, but more important is to know what it means, what this ordinance includes and I know more or less what it is trying to regulate, but with respect to dogs, I don't know if other types of animals or other types of pets, the most common are dogs and cats, but there are an infinite number of pets, but even so I don't know of any type of regulation for cats.

4.1.2.3.   According to the documentary analysis guide of the regulations issued by the Provincial Municipality of San Martín.

The only ordinance enacted by the Provincial Municipality of San Martín was analysed, with the following result:

Municipal Ordinance N°014-2020-MPSM

Municipal Ordinance N° 014-2020- MPSM, establishes the regulations for dog ownership and registration in the district of Tarapoto. From all the articles analysed it can be inferred that:

1. The regulation consists of 36 articles, which only cover the keeping and registration of dogs, and does not cover any other domestic animals.

2. Within the dog ownership and registration regulations, we can identify that the authority tries to provide welfare and protection to dogs, indicating how they should be cared for and kept; it also establishes sanctions for owners and/or keepers in case of non-compliance (see articles 28, 32, 33, 34, 35 and 36 of the documentary analysis guide of Ordinance 014-2020-MPSM).

3. It indicates that the training must be carried out in an authorised centre and must not be aimed at strengthening the dog's aggression, and it is strictly forbidden for these centres to organise dog fights.

4. Dogs must be marketed in authorised places and the seller must provide the buyer with all information concerning the breed of the dog.

5. The registration of potentially dangerous dogs must be carried out within a maximum

period of three months from birth.

6.   It establishes the signing of inter-institutional agreements with public or private animal welfare institutions to shelter abandoned dogs.

7.   Guide dogs are allowed to travel on any means of transport, as well as free admission to establishments, places and public performances.

8.   It provides that in the case of adoption of very dangerous or potentially dangerous dogs, owners must have a positive psychological evaluation certificate.

9.   In the situation where a dog bites or injures a person or animal, the dog shall be kept under observation and in case of serious injury or death, the dog shall be put down, except for a dog that acted in defence of its owner, itself or its offspring.

10. Minor offences are fined up to 0.5 UIT, serious offences up to 1 UIT and very serious offences up to 2 UIT. The severity of the sanction depends on the danger caused to society, whether the offence is repeated and the profit generated by committing the offence.

4.1.3. According to Specific Objective 3

Specific Objective N° 3: To identify compliance with the policies implemented by the Provincial Municipality of San Martín in the district of Tarapoto with regard to domestic animals.

4.1.3.1.   Results of the survey applied to the problems of the district of Tarapoto.

From the survey carried out, the question aimed at answering this objective is the question 4, which yields the following results.

4. Do you think that the Provincial Municipality of San Martin is applying any policy in the district of Tarapoto in favour of the welfare and protection of domestic animals?

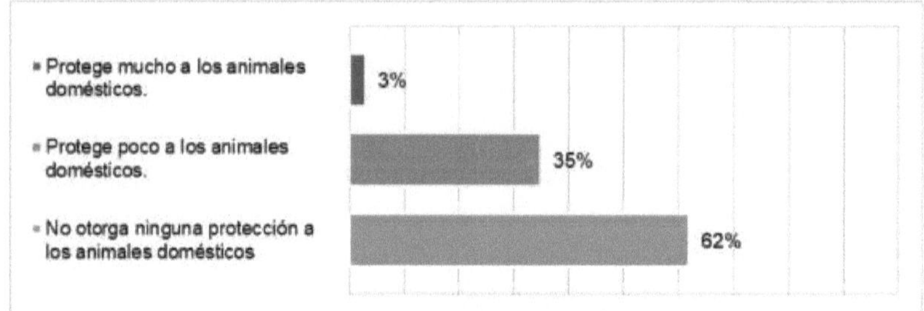

Figure 4

Do you think that the Provincial Municipality of San Martin is applying any policy in the district of Tarapoto in favour of the welfare and protection of domestic animals?

Source: Own elaboration

From Figure 4 we can see that 62% of the surveyed population does not believe that the Municipality is applying any policy in favour of domestic animals in the district of Tarapoto, while 38% indicate that the Municipality is applying some policy in favour of domestic animals.

4.1.3.2. Results of the interview with experts in the field of protection and welfare of

pets

The question aimed at answering this objective is question 5, with the following result.

5. Is the Provincial Municipality of San Martin applying any policy in the district of Tarapoto in favour of the welfare and protection of domestic animals?

Table 13

Question 5.

Expert response	
Marco Antonio Sánchez Huaripata (Deputy Manager of Local Economic Development of the MPSM)	**Elizabeth García Panduro (Representative of the non-profit association Yupi Wasi)**
In fact, we have been carrying out an inspection activity based on the ordinance that we have and the conditions that these pets must have. Previously, we have also had quite a few complaints about improper breeding of pets or animal abuse, which is also covered by the ordinance. We are constantly dealing with complaints about dog bites as well, which is where the ordinance comes in. We are at a time of socialisation with the population so that they already have the corresponding permits to keep pets; what our inspection personnel do is to see the conditions in which these pets are being kept, it is hard work, the city is large and sometimes it is a little difficult for the population to provide conditions for pets, and that is what makes our work difficult; What we do not want to see is dogs that are abandoned on public roads, or kittens, or other types of pets that we abandon due to lack of space or lack of conditions, that is also what the ordinance regulates.	I have not heard anything about this, either through us or elsewhere, but I do know that at project level, they are looking at the issue of seeking an agreement with the aim of creating awareness among people, to provide better quality of life and responsible pet ownership, I heard that the Municipality is trying to put together a plan, but it is always very slow, it is creating agreements with the University, with the Veterinary College and I do not know if with other private veterinarians who can get involved to create, for example, a clinic for basic care, obviously for dogs; at least there is a project for people with limited resources who can't afford a vet, but who want their dog to be looked after. I know it's a project, I know they are talking, trying to get more professionals involved, like veterinary students. I also know that they don't want to make a shelter because people don't have the awareness of responsibility and it would be like "there is a place where I can leave this dog, so I go and let it go", that's one reason why not a shelter.

4.1.4. According to Specific Objective 4

Specific Objective N° 4: To establish the relationship between the protection and welfare of domestic animals and the policies implemented by the Provincial Municipality of San Martin in the district of Tarapoto, 2020.

4.1.4.1. Results of the survey applied to the problems of the district of Tarapoto

The questions aimed at answering this objective are questions 5, 6, 7 and 8, which yield the following results.

5. What level of protection do you consider that domestic animals are afforded by the Provincial Municipality of San Martín?

Figure 5

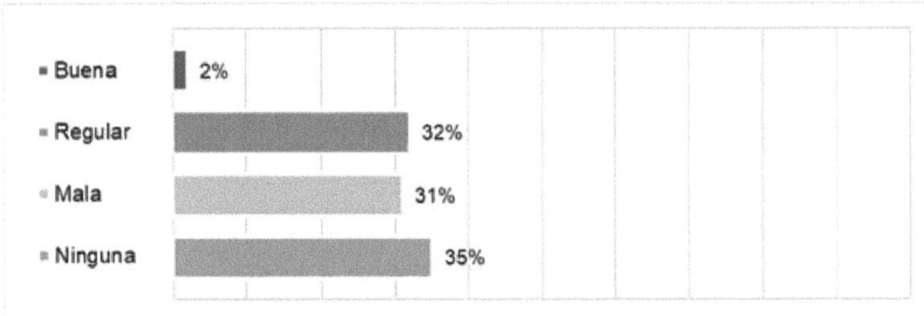

What level of protection do you consider that domestic animals are afforded by the Provincial Municipality of San Martín?

From Figure 5 we can see that 31% of the surveyed population considers that the Municipality grants "bad" protection to domestic animals, while 32% indicate that the Municipality grants "fair" protection to domestic animals, as well as 2% indicate that the Municipality grants "good" protection to domestic animals and 35% say that the Municipality grants "no" protection to domestic animals.

6.   Do you know if a municipal animal shelter has been created in the city of Tarapoto?

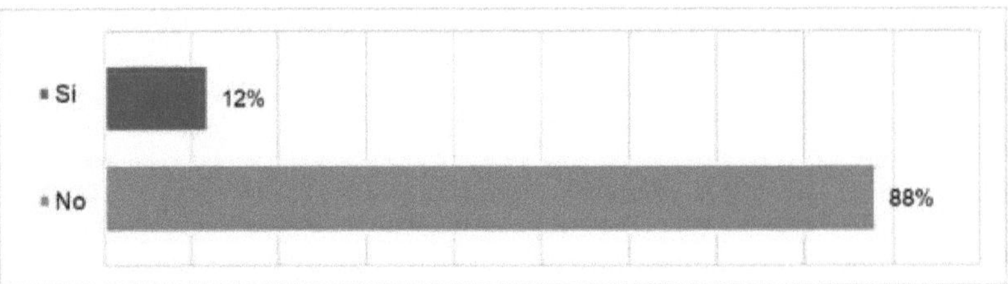

Figure 6

Do you know if a municipal animal shelter has been created in the city of Tarapoto?

Source: Own elaboration

Figure 6 shows that 88% of the population surveyed did not know that a municipal animal shelter had been created in the city of Tarapoto, while 12% stated that they did know that a municipal animal shelter had been created in the city of Tarapoto.

7.   Do you have any knowledge of whether the city of Tarapoto is giving the public awareness campaigns on domestic animals?

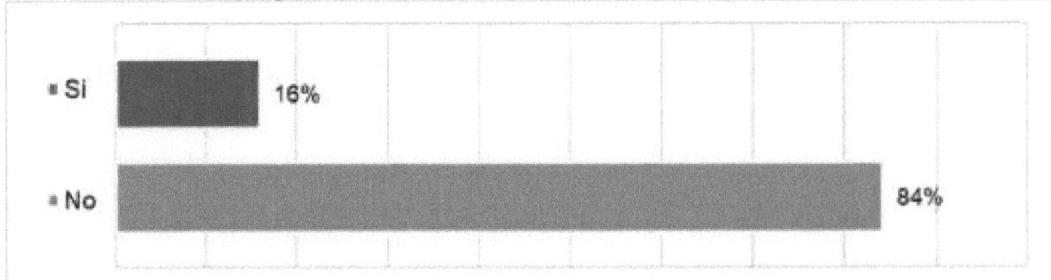

Figure 7

Do you know if awareness campaigns are being carried out in the city of Tarapoto regarding domestic animals?

Figure 7 shows that 84% of the surveyed population is not aware that awareness-raising campaigns are being carried out in the city of Tarapoto, while 16% say that they are aware that awareness-raising campaigns are being carried out in the city of Tarapoto.

8.   Do you consider that in the district of Tarapoto, the Municipality is taking any action to positive action in favour of the situation of stray domestic animals?

Figure 8

Do you think that in the district of Tarapoto the Municipality is taking any positive action to help the situation of stray pets?

Source: Own elaboration

In Figure 8 we can see that 94% of the surveyed population considers that the Municipality is not taking any positive action regarding the situation of stray domestic animals, while 6% indicate that the Municipality is taking positive action regarding the situation of stray domestic animals.

9.   Did you know that pet abuse can be reported?

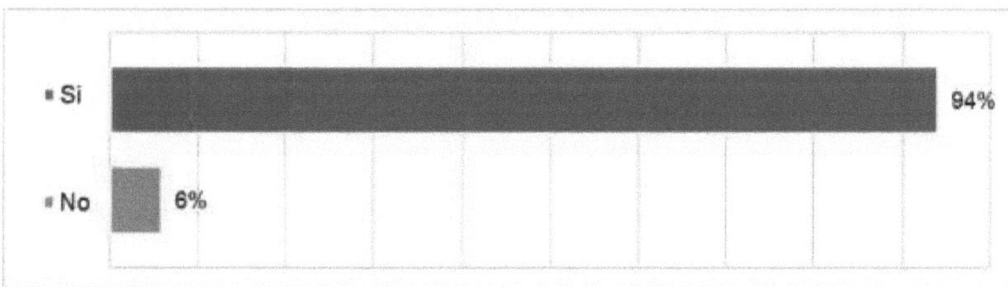

Figure 9
Did you know that pet abuse can be reported?

In Figure 9 we can see that 94% of the population surveyed said that they are aware that domestic animal abuse can be reported, while 6% indicated that they do not know that domestic animal abuse can be reported.

10. Have you ever reported this type of abuse?
Figure 10

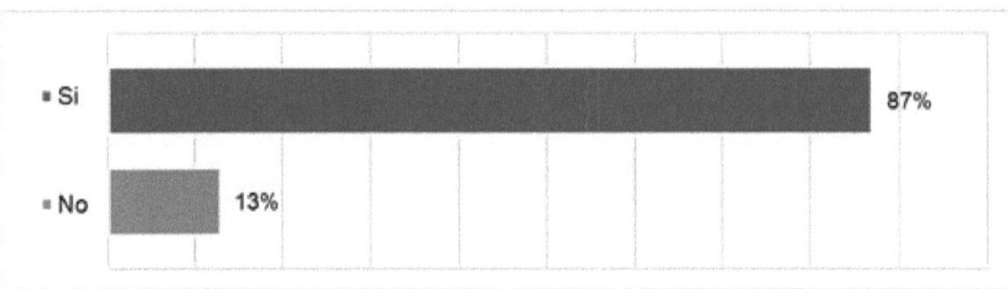

Have you ever reported this type of abuse?

Source: Own elaboration

In Figure 10 we can see that 87% of the population surveyed stated that they had reported pet abuse, while 13% stated that they had not reported pet abuse.

4.1.4.2. Results of the interview with experts in the field of protection and welfare of pets

The questions aimed at answering this objective are questions 6,7,8, with the following result.

6. What is the position of the Provincial Municipality of San Martin regarding the creation of an animal shelter?

Table 14

Question 6.

Response from expert	
Marco Antonio Sánchez Huaripata (Sub-Gerente de Desarrollo Económico Development of the MPSM)	**Elizabeth García Panduro (Representative of the non-profit association Yupi Wasi)**
We would like to have a municipal dog shelter or a municipal pet shelter, but at the moment circumstances still prevent us from doing so, due to logistical issues where the economic issue has a lot of influence, this has limited us; We are working, we have some conversations with the animal protection associations, with them we are going to work together to socialise this whole issue and start rescuing some dogs abandoned in the street, we are not going to do it directly as a municipality, but through the associations and that will also be a great relief for some pets that live in conditions of neglect.	I know that it is not in the plans of the Municipality, because if people know that there is a shelter, they would abandon their animals or it has happened with us, that we are not a shelter, but they know that we rescue animals, and they have come to leave their pets with us. People often think that this is limited; we are working, we have some conversations with the animal protection associations, with them we are going to work together to socialise this whole issue and start rescuing some abandoned dogs in the street, we are not going to do it directly as a municipality, but through the associations and that will also be a great relief for some pets that live in conditions of neglect.

7. Is the Provincial Municipality of San Martin   currently carrying out the following

activities
awareness-raising campaigns?

Table 15
Question 7.

Response from expert	
Marco Antonio Sánchez Huaripata (Sub-Gerente de Desarrollo Económico Local MPSM)	**Elizabeth García Panduro (Representative of the non-profit association Yupi Wasi)**
Yes, based on the ordinance we have, we have already been carrying out some socialisation activities related to responsible pet ownership; If you check the Municipality's website you will see that we are continually publishing some notices on responsible pet ownership, one of them is that if you take your pet for a walk, don't forget to take it out with its muzzle, its leash and above all don't forget to bring your waste bags, because you have to bear in mind that the same green spaces that our pets occupy are also occupied by our children and that is something that could work against us, but that is why owners are urged to bring their bags so that they can pick up the waste; The Municipality's web page is constantly being communicated.	Not yet, I have the hypothesis that it does have the idea of doing so, as the agreements are being worked on so that the associations that work in rescue get involved in going and giving talks, raising awareness and also the professionals. When there is a campaign, vaccination, cleaning, there should also be a stand to touch on this subject, to raise awareness, why not see an animal as a gift that you are only going to have for a while, the fact of also giving a chance to dogs that have a disability or that are already old, I think that giving them that quality of life has to do with human quality as well, as well as being aware of how many animals you can have, because it involves money, space and time.

8. What positive action is the Provincial Municipality of San Martin taking to help the situation of domestic animals?

Table 16
Question 8.

Response from expert	
Marco Antonio Sánchez Huaripata (Sub-Gerente de Desarrollo Económico Development of the MPSM)	**Elizabeth García Panduro (Representative of the non-profit association Yupi Wasi)**
This is not only the action, we have been carrying out a series of activities in favour of our pets in Sint Maarten, among them is the registration, we are doing a national registration and an international registration, through a certification, this is going to be granted to each of the pets, what does this mean? This means that if I take my dog on a trip for A or B, I have to obtain an authorisation and the authorisation includes a series of requirements; international registration allows me to take my dog abroad without any problem, as well as placing a microchip where all its data and history will be registered. The sanitary control implies that I will have to present their health card, that they have their vaccinations, their deworming, their medical check-up and this will be included in the record that my pet will have within the Municipality. In reference to this project, we	Well, I think that at least this issue of seeking to make agreements is a first step, and also the fact that I know that Rosa, who is the first to start this association, has been called to talk, they have called her to tell her of their intentions, so I think that this is also a positive attitude, of wanting to be in contact with the entities or people who can help to promote animal protection and responsible ownership, I see that at least the Municipality is looking to create spaces, perhaps it has many gaps, yes definitely, and I think it is more because of the bureaucracy that it really is, but at least they are making this movement that we hope will become more fluid and we will be able to see and know more things, not only the associations or shelters, but also the population in general, know what to do and how to get involved, I hope it will be in the shortest possible time.

were talking about an urban nucleus considering Tarapoto, Morales and La Banda de Shilcayo, also from Juan Guerra to Cacatachi,	

but for the moment we have been working with the Municipality of Morales, which already has its ordinance as well; With regard to responsible pet ownership and with La Banda de Shilcayo, I believe that their regulation is on its way, this makes it easier for us to work together and carry out much more efficient actions. We also work together with the OGESS - Bajo Mayo, which is leading a very good work plan, with the Professional Association of Veterinary Doctors in San Martín, where we have an agreement and where the first Municipal Veterinary Clinic is going to be built, for logistical reasons the Veterinary College will build and manage it, and also, we have an agreement with the School of Veterinary Medicine - UNSM, the school has already built a Veterinary Hospital that would help people who need a veterinary service at a social cost. We have to take into account that health costs in general are high, both for people and for pets, based on this, we have noticed the need to create these services through agreements and conversations that we already have with the institutions. We also support the OGESS - Bajo Mayo, especially in the vaccination campaign against canine rabies, which is carried out every year here in the city of Tarapoto. In addition, taking into account that the University has an area of social projection, which will also help us to provide a more affordable service to the public, so that all pets can access a veterinary service or a consultation for a medical check-up if necessary.

4.2. Discussion

Regarding the discussion of the results:

4.2.1. According to the specific result 1

In order to respond to specific objective 1, the documentary analysis guide and the survey were used as instruments. The results of the documentary analysis guide show that there are 26 articles directly related to the subject matter of the research, 6 of which prescribe the role of local governments in relation to animal protection and welfare, They have the duty to deal with complaints and ensure that the law is applied, to encourage the creation of shelters, to grant authorisations and licences in cases of transport, ownership, commercialisation and veterinary care, to be an active part of the animal protection committee, to stop animal abandonment and to impose sanctions.

Comparing the content of the law N°30407 regarding domestic animals with the theories proposed by the philosophers Peter Singer and Tom Regan, it can be determined that this law is not framed within the philosophical positions that these authors propose, since both Peter Singer and Tom Regan consider in their theories that animals have the same moral status and rights as human beings, This is contrary to the aforementioned regulations because, in our country, animals are not considered as subjects of law but as part of the patrimony of human beings, and therefore, the conceptions established in the law are not equivalent to rights for animals, but are regulated to establish a social order; Similarly, this coincides with the findings of Moron (2019) in his undergraduate thesis, where he concluded that there is a controversy between Law No. 30407 - Animal Protection and Welfare Law with Article 206-A on Abandonment and cruel acts against domestic and wild animals of the Penal Code, because the legislation qualifies animals as sentient beings and all the regulations in Peru offer protection to domestic animals, but, considering them as part of the patrimony of human beings, leaving a controversy as to whether they are really protecting animal life or only the economic purpose of the animal.

Likewise, as a result of the results obtained in the survey, it is identified that 62% of the population of the district of Tarapoto indicates that they are aware of Law No. 30407 and its articles on domestic animals, while 38% state that they do not know, so, comparing these results, we conclude that Law No. 30407 and its articles on the welfare and protection of domestic animals are known by most of the inhabitants of the district of Tarapoto; This is contrary to the results obtained by Butriago (2019) in his degree thesis, who obtained unfavourable results, concluding that, in the absence of knowledge of the population about animal protection laws, the environment and animal welfare are in danger. Contrasting the results obtained in this research with those obtained by Butriago, we state that, in our case, it is favourable for the welfare and protection of domestic animals that the inhabitants of our district are aware of the law and its articles.

4.2.2. According to the specific result 2

In order to respond to specific objective 2, the documentary analysis guide and the survey were used as instruments. With regard to the regulations issued by the Provincial Municipality of San Martín, we have Municipal Ordinance N° 014-2020 -MPSM, which was analysed through the documentary analysis guide, obtaining as a result that, in its 36 articles, it only regulates the ownership and registration of dogs.

It should also be noted that Law N°30407 - Law for the Protection and Welfare of Animals was enacted in 2015, however, the investigation showed that five years later, in 2020, the Provincial Municipality of San Martin only issued a municipal ordinance in favour of domestic animals, taking into account only dogs, showing the little importance that the authorities have given to the issue of protection and welfare of domestic animals in the city of Tarapoto.

According to the theories of the philosophers Singer and Regan, all animals have a moral status because they are sentient beings and therefore animals have rights, but this is not directly related to the municipal ordinance, since this regulation only regulates domestic animals "dogs", but not other domestic animals; furthermore, it is inferred that this municipal ordinance is established without considering, as well as the law, animals as subjects of law.

From all that has been analysed we can see that the regulations issued by the Provincial Municipality of San Martin are deficient when talking about the Protection and Welfare of Domestic Animals, since Law N°30407 does not only consider dogs within this classification, it should be noted that the law classifies animals into domestic and wild animals; Likewise, this partially coincides with what was obtained by Luque (2019) in his undergraduate thesis, where he determined that the Peruvian State did enact regulations such as Law N°30407 to protect domestic animals, however, despite this, in Arequipa as well as in its districts there is a lack of public regulatory acts implemented by local governments, resulting in the ineffectiveness of the aforementioned law for its application in the region and thus affecting both animals and public health.

In reference to the survey applied, 60% of the inhabitants of the district of Tarapoto are aware of any decree or ordinance issued by the authorities of the Provincial Municipality of San Martín in favour of domestic animals, so this result shows that the majority of the population is aware of the municipal ordinance N° 014-2020-MPSM; However, it is not enough to know, but there must also be compliance by citizens, so, contrasting with what was obtained by Chú (2016) in his undergraduate thesis, who identified that in the Punta del Este sector there is animal abuse, so it would be necessary to identify whether with the implementation of the municipal ordinance in 2020, there is any improvement over the results obtained by Chú.

4.2.3. According to the specific result 3

In order to respond to specific objective 3, the survey and the interview guide were used as an instrument to interview the experts, obtaining as a result of the survey that 62% of the inhabitants of the district of Tarapoto consider that no policy in favour of Animal Protection and Welfare is being applied. Likewise, as citizens of the city of Tarapoto we are witnesses that this ordinance is not effectively applied, since we were able to observe, during the data collection of the investigation, in a direct way, the lack of action or promotion of any activity in benefit of domestic animals, as well as the lack of a specific area within the Municipality in charge of the registration of dogs, as stated by the official of the Municipality in the interview, contravening the stipulations of the Municipal Ordinance 014-2020- MPSM.

Contrasting with the answers obtained in the interview with the experts on the Protection and Welfare of Domestic Animals, we can deduce in relation to the application and enforcement of public policies in the city of Tarapoto that there is little compliance on the part of the Municipality because, of all that is stated in the ordinance, both experts agree that only the municipal authority is acting in response to reports of animal abuse, leaving aside everything else stipulated in the ordinance. It should be mentioned that in our district, the municipal authority is far from issuing any regulations that fall within the parameters of the positions of the philosophers Tom Regan and Peter Singer.

4.2.4. According to the specific result 4

In order to respond to specific objective 4, the survey, the documentary analysis guide and the interview guide with experts in the protection and welfare of domestic animals were used as instruments. The results of the survey showed that 35% of the inhabitants of the district of Tarapoto stated that the municipality does not provide any protection to domestic animals, Of the 300 inhabitants surveyed, only 5 responded that the municipality provides good protection for domestic animals, which means that the majority of the inhabitants consider that the municipality does not provide any protection for domestic animals.

It was also found that 88% of the population surveyed indicated that they were not aware that a municipal animal shelter had been created, and although 12% said they were aware, the reality is that, to date, there is no municipal animal shelter in the district of Tarapoto; This position of the population is reinforced by the representative of the animal shelter "Yupi Wasi", who stated that there is no concrete agreement regarding the support of the Municipality for animal shelters or the creation of a municipal shelter, in contravention of Article 8 of Law N°30407 on the creation of temporary shelters.

Likewise, 84% of those surveyed indicated that they were not aware that awareness campaigns were being carried out with regard to domestic animals in the district of Tarapoto, while 16% stated that they were, which means that compliance with the law and the ordinance by the population is scarce.

Furthermore, 94% of the population surveyed said that the Provincial Municipality of San Martín is not taking any positive action to help the welfare of stray domestic animals, while 6% said it was, so it can be deduced that the Municipality, in the eyes of the majority of the population, is not taking any action to help domestic animals, in contravention of Law N°30407 and its own ordinance N°014-2020-MPSM.

It also turned out that 94% of the inhabitants who participated in the survey know that domestic animal abuse can be denounced, while 6% said that they do not, which suggests that the majority of the population is aware that animal abuse is a crime.

Finally, 87% of the population surveyed in the district of Tarapoto stated that they had never reported animal abuse, while 13% stated that they had, so it can be inferred that the population of Tarapoto does not have a culture of reporting domestic animal abuse.

Contrasting with the documentary analysis guides, the Law for the Protection and Welfare of Animals - Law N°30407 as well as the Municipal Ordinance N°014-2020-MPSM, with the survey carried out with the inhabitants of the district of Tarapoto, it can be inferred that, The regulations issued by the Provincial Municipality of San Martin in the district of Tarapoto in favour of domestic animals are scarce and are not properly implemented, so there is very little knowledge on the part of the population with regard to the regulations that protect domestic animals in the city of Tarapoto, This in turn is consistent with the fact that the authority responsible for this matter, the Provincial Municipality of San Martin, is not carrying out the corresponding actions to ensure that these regulations are complied with and as a result, most of the population does not take positive actions for the benefit of domestic animals, which coincides with what Cavalcanti (2016) pointed out, since the population does not know the regulations, it is incompetent, causing that even today, animals continue to be victims of the ignorance of society.

In reference to the obtained and comparing it with the theories of the philosophers Peter Singer and Tom Regan we can deduce that, first, both the Law N°30407 and the municipal ordinance 014-2020-MPSM, have been regulated considering domestic animals as part of the property of the human being, second, these philosophers have the position that animals are entities that enjoy a moral status equivalent to the human being, secondly, these philosophers have the position that animals are entities that enjoy a moral status equivalent to that of human beings, therefore, for them animals also enjoy rights, which must be respected, which is why we infer that these two regulations are not framed within these theories, which can be seen throughout the law and the ordinance.

With regard to the analysis of the interviews with the actors involved in the subject of Protection and Welfare of Domestic Animals, it could be inferred that the Municipality has projects that favour domestic animals that are not yet executed, since they are in development and with the objective of being implemented in the city of Tarapoto, likewise, the intention of the Municipality to establish agreements with the different institutions involved in the subject of Protection and Welfare of domestic animals is visible, It is thus from the perspective of the non-profit association "Yupi Wasi", that it knows in part of the intentions and projects to be developed by the Municipality, since it communicated personally with the representative of the association, leaving also in evidence that for the moment it is not applying the current regulations issued by the Municipal authority but that they are coordinating and working the issue so that soon they can provide Protection and Welfare of Domestic Animals.

All animals must be protected by the State, through the decentralised bodies that are in charge of enforcing the law through policies, such as the Provincial Municipalities according to their area of competence. This regulation is necessary because it allows coexistence in harmony among all living beings; the protection and welfare of domestic animals has positive effects on the daily life of man, as it allows the maintenance of health and public order; coinciding with what was stated by Burgos (2019) who stated that all levels of government must implement public policies including companies, citizens, organisations, institutions, etc., to contribute to public health and the promotion of new policies.

4.2.5. Discussion of the general objective

Although our general hypothesis is that the relationship between the protection and welfare of domestic animals and the policies implemented by the Provincial Municipality of San Martin in the district of Tarapoto, 2020 is positive; according to the survey, expert interviews, background information, theories and documentary analysis guides (analysed in the results of specific objectives 1, 2, 3 and 4), the Provincial Municipality of San Martin has neither implemented nor complied with the Law for the Protection and Welfare of Animals - Law N° 30407 and the same situation can be inferred with regard to Municipal Ordinance N°014-2020-MPSM. Therefore, it can be observed that almost all the citizens of the district of Tarapoto are unaware of these regulations, and consequently, there are no positive actions that support the application of the aforementioned regulations in the locality, resulting in a negative relationship.

CONCLUSIONS

1.   It was identified that the Animal Protection and Welfare Law - Law N°30407 has 26 articles, where only 6 of them prescribe the role and participation of local governments in the application of the law; on the other hand, regarding the knowledge of the population of the district of Tarapoto about this law, it was concluded that the majority of the inhabitants belonging to the district of Tarapoto are aware of the existence of Law N°30407.

2.   It was identified that the Provincial Municipality of San Martin has only implemented the Municipal Ordinance 014-2020-MPSM, which approves the regulation on the regime of dog ownership and registration in the district of Tarapoto, apart from that there is no other policy implemented by the Municipality, also, it was found that most of the population of Tarapoto is not aware of the application of this ordinance, as well as other policies by the municipal authority.

3.   It was identified that in the city of Tarapoto only the Municipal Ordinance 014-2020-MPSM, regulations applied to canines, is regulated and it was concluded that this ordinance is not being effectively enforced by the Municipality in our district to date.

4.   It was established that there is a deficient relationship between the protection and welfare of domestic animals and the policies implemented by the Provincial Municipality of San Martin in the district of Tarapoto in 2020, with the provisions of Law No. 30407, because the regulation issued by the Municipality does not currently fall within the parameters of the latter, since it only considers dogs as domestic animals and, in addition to this, the municipal ordinance is not currently being implemented.

5.   From all this it was determined that there is a negative relationship between the protection and welfare of domestic animals and the policies implemented by the Provincial Municipality of San Martin in the district of Tarapoto in 2020, which is contrary to our initial hypothesis; In addition, it was determined that this relationship is negative because the Provincial Municipality of San Martin has only one ordinance which is neither applied nor effectively implemented, resulting in the majority of the population of the district of Tarapoto not being aware of the regulations in the city of Tarapoto, and not taking positive actions that contribute to the protection and welfare of domestic animals.

RECOMMENDATIONS

Taking into account the above and the results we have obtained with regard to the research topic, we recommend the following for the benefit of the protection and welfare of domestic animals:

1.   The Provincial Municipality of San Martin must correctly implement the ordinance issued, as it is of no use to have a regulation if it does not make the effort to apply it, as well as to extend this ordinance without limiting it only to dogs.

2.   The Provincial Municipality of San Martin should prioritise awareness-raising regarding animal protection and welfare, so that the population knows how to act in cases of animal ownership, cruelty and mistreatment. In addition to this, it should inform about its attributions and its intervention in these cases.

3.   The Provincial Municipality of San Martin should encourage the population to sterilise their pets, as this contributes to the reduction of animal abandonment and neglect.

4.   It is also recommended that the Provincial Municipality of San Martin create a centre for primary medical care for pets, where people with limited resources can go with their pets when they require urgent attention.

5.   The Provincial Municipality of San Martin should continuously implement deworming campaigns for domestic animals, either free of charge or at social cost, in order to contribute to the health of the animals, their owners and society as a whole.

6.   The Provincial Municipality of San Martin and the population of the district of Tarapoto should not take animal abuse for granted, but on the contrary, it should be analysed in all possible areas, as aggressive behaviour towards any living being is not normal and becomes a warning sign that could become a major problem in the future.

BIBLIOGRAPHICAL REFERENCES

Acero, M. (2017). The Human-Companion Animal Relationship as a sociocultural phenomenon. Perspectives for public health (Thesis for the degree of Doctor of Public Health). National University of Colombia, Bogotá.

Burgos, J. A. (2019). Lineamientos para la política pública de bienestar animal en el Municipio de Sibaté (Cundinamarca) Colombia. Colombia: Pontificia Universidad Javeriana. https://repository.javeriana.edu.co/bitstream/handle/10554/46836/tesis%202020 %20final%2023%20de%20enero%20de%202020.pdf?sequence=2&isAllowed=y

Butriago, O. (2019). Promotion and communication in animal welfare and responsible pet ownership in Colombia. Medellin, Colombia: UNIVERSIDADCES . Obtained from https://repository.ces.edu.co/bitstream/handle/10946/4706/Tesis%20de%20grad o?sequence=2&isAllowed=y

Castillo, J. L. (2015). "Eficacia de la aplicación de la normativa en los casos de abandono y actos de crueldad contra animales silvestres en la región Lambayeque periodo 2015. Pimentel: Repositorio de la Universidad Señor de Sipán. Retrieved from https://repositorio.uss.edu.pe/bitstream/handle/20.500.12802/5215/Castillo%20C arri%C3%B3n%20Jos%C3%A9%20Luis.pdf?sequence=1

Cavalcanti, G. (2016). Abandonment of domestic companion animals, followed by death and criminal liability under Law No. 30407 in Peru (Thesis for the professional title of lawyer). Alas Peruanas University, Lima.

Cervantes, W. S., & Huiman Jimenez, M. C. (2021). La implementación de niveles de justicia hacia el maltrato animal en el Perú, 2020. Lima: Repositorio de la Universidad César Vallejo. César Vallejo. https://repositorio.ucv.edu.pe/bitstream/handle/20.500.12692/73682/BulnesCW S-Huiman JMC-SD.pdf?sequence=1&isAllowed=y

Chú, A. C. (2016). Diagnóstico sobre tenencia responsable de animales de compañía en trece sectores del distrito de Tarapoto - Provincia de San Martín. Tarapoto: Universidad Nacional de San Martín. Retrieved from https://repositorio.unsm.edu.pe/bitstream/11458/1087/1/TP MVE 00003 2016.p df

Cruz, S. (26 April 2019). Legis Ámbito Jurídico. Retrieved from Animals as family members, is regulation necessary: https://www.ambitojuridico.com/noticias/informe/civil-y-familia/animales-como- miembros-de-la-familia-es-necesaria-una-regulacion.

dpej.rae.es. (2023). Pan-Hispanic dictionary of legal Spanish. Retrieved from https://dpej.rae.es/lema/protecci%C3%B3n-de-los-animals#:~:text=En%20general%2C%20conjunto%20de%20normas,y%20sacrif icio%20de%20los%20animales.

Dzul, M. (2010). Unit 3. Basic application of scientific methods. Non-experimental design. https://www.uaeh.edu.mx/docencia/VI Presentationsflicenciaturaenmercadot ecnia/fundamentosdemetodologiainvestigacion/PRES38.pdf

The Provincial Council of San Martín (2020, August 14). Municipal Ordinance N°014- 2020-MPSM. Tarapoto: Provincial Municipality of San Martín. Retrieved from https://cdn.www.gob.pe/uploads/document/file/2476161/Ordenanza%20Municipal %20N%C2%B0%20014-2020-AflMPSM.pdf

The Congress of the Republic (1997, 15 July). General Health Law. Diario Oficial El Peruano.

http://www.essalud.gob.pe/transparencia/pdf/publicacion/ley26842.pdf

The Congress of the Republic (2003, 26 May). Ley Órganica de Municipalidades. Lima: Diario Oficial El Peruano. Retrieved from

https://diariooficial.elperuano.pe/pdf/0015/3-ley-organica-de-municipalidades-1.pdf?fbclid=IwAR3WVAQgkq4e9leRraeKP5o4ZNRo2j-kMOZGnR1fUKtBG7f1RwnlORsDQsl

The Congress of the Republic (2003, 26 May). Law N° 27972, Organic Law of Municipalities. Lima: Diario Oficial El Peruano. Retrieved from

https://www.mef.gob.pe/contenidos/presupubl/capacita/programacionformulaci onprespresupuestal2012/Anexos/ley27972.pdf

The Congress of the Republic (2016, 07 January). Animal Protection and Welfare Law N°30107. Lima: Diario El Peruano. Retrieved from

https://busquedas.elperuano.pe/normaslegales/ley-de-proteccion-y-bienestar- animal-ley-n-30407-1331474- 1/?fbclid=IwAR0vpjlZ9Hanmxl2tTE 97lasaJ01JDKmA6OS-6fY-m345sCLi7qG zRxTg

The Congress of the Republic (2016, 07 January). Law No. 30407 - Law on Animal Protection and Welfare. Diario Oficial El Peruano. Retrieved from https://busquedas.elperuano.pe/normaslegales/ley-de-proteccion-y-bienestar- animal-ley-n-30407-1331474-1/

The Government of Emergency and National Reconstruction (1992, 29 December). Decree Law No. 26162 - Law of the National Control System. Lima: Diario Oficial ElPeruano   . obtained   from

http://doc.contraloria.gob.pe/documentos/Ley%20del%20Sistema%20Nacional% 20of%20Control-Decreto%20Ley%20N%C2%B0%2026162,%20publicada%20el%2030.Dic.1992 x.pdf

Fernández, P., & Días, P. (May 27, 2002). Quantitative and qualitative research. Spain. https://ocw.unican.es/pluginfile.php/355/course/section/154/Tema%25208.pdf

Fiedrich, N. O. (2012). Animal Welfare. Sitio Argentino de Producción Animal. Retrieved from https://www.produccion-animal.com.ar/etologiaybienestar/bienestar engeneral/32- Bienestar Animal.pdf

Animal Protection and Welfare Law, LAW N° 30407 (Congress of the Republic December 16,

2015).

López, N. F. (2019). Analysis of the norms that sanction the mistreatment of domestic animals in relation to dogs and cats, according to Peruvian legislation Nuevo Chimbote 2015. Bachelor's thesis, Universidad César Vallejo, Chimbote.

Luque, K. (2019). Analysis of the situation of companion animals in a state of abandonment and the implementation of a municipal legal regime of protection in the district of Paucarpata, Arequipa, 2018-2019. (Thesis to obtain the professional title of lawyer). San Francisco Autonomous University, Arequipa.

Moron, M. (2019). La controversia del delito de Abandono y Actos de Crueldad contra Animales Domésticos y Silvestres tipificado en el Art. 206-A del Código Penal con la Ley N°30407, respecto a su calificación como seres sensibles. (Thesis to obtain the professional title of lawyer). Andean University of Cusco, Cusco.

MPSM (26 January 2022). Opening Institutional Budget 2022. Retrieved from TransparencyPortalMPSM :

https://mpsm.gob.pe/public/uploads/documentos/PIA 2022 MPSM.pdf

O, S. V., & W., R. W. (2016). Analysis of Law 30407 "Animal Protection and Welfare Law" in Peru. Scielo Peru. Retrieved from http://www.scielo.org.pe/scielo.php?script=sci arttext&pid=S1609- 91172016000200023&lang=en

World Health Organisation. (1946, 22 July). Constitution of the World Health Organization. New York: Official Journal, volume LVI, number 32, 9 September 1949 .

https://www3.paho.org/gut/dmdocuments/Constituci%C3%B3n%20de%20la%20 World%20Health%20Organisation%20of%20Health%20.pdf

Ortega Peñafiel, S. A., Maldonado Cabrera, M. D., Bejarano Paz, L. G. and Freire Goyes, V. E. (2021). Infractions, penalties and fines for animal abuse in Latin America. SOCIALIUM, 16. Obtained from

https://revistas.uncp.edu.pe/index.php/socialium/article/view/815/1056

Pérez, J. A. (2007). Las variables en el método científico. Scielo Perú, 5. Retrieved from http://www.scielo.org.pe/scielo.php?script=sci arttext&pid=S1810- 634X2007000300007

Piamore, E. (21 October 2019). Animal Expert. Retrieved from Animal abuse - Types, causes and how to report: https://www.expertoanimal.com/el-maltrato- animal-types-causes-and-how-to-report-

24291.html?fbclid=IwAR0VRykp8DRP4UMxuyfaATJbhO8pmimChSyJh7UYZ7h dOTFwtoJ68qC8nA

Quinto Juzgado Penal Unipersonal Supraprovincial de Chiclayo y Ferreñafe (2021, July 6th). Sentence. Chiclayo: Superior Court of Justice of Lambayeque. Obtenido de https://img.lpderecho.pe/wp-content/uploads/2021/07/Expediente- 06261-2020-11-1706-JR-PE-01-LP.pdf

Ruiz, D., & Cardénas, C. (2016). IUS Revista Jurídica. Retrieved from What is a public policy: https://www2.congreso.gob.pe/sicr/cendocbib/con4uibd.nsf/8122BC01AACC9C 6505257E3400731431/%24FILE/QU%C3%89ES UNA POL%C3%8DTICA P %C3%9ABLICA.pdf.

Rus, E. (01 November 2020). Economipedia. Retrieved from Correlationai Research: https://economipedia.com/definiciones/investigacion-correlational.html

Schunemann, A. (June 2011). Scielo. Retrieved from Animal Welfare in Veterinary Medicine and Zootechnics Education: Why and what for: https://www.scielo.org.mx/scielo.php?script=sci arttext&pid=S0301- 50922011000200004

SENTENCE, 01413-2017 (Constitutional Court 2018). Retrieved from https://tc.gob.pe/jurisprudencia/2019/01413-2017-AA.pdf

Suarez, L. (2021). Las políticas públicas y la toma de decisiones en el Perú. Universidad Continental. Retrieved from https://blogposgrado.ucontinental.edu.pe/las-politicas- public-policies-and-decision-making-in-peru.

Tarazona, D. (23 April 2020). Animal abuse: Inhumanity visible through social networks. Retrieved from Punto Seguido - UPC: https://puntoseguido.upc.edu.pe/el-maltrato-animal-la-inhumanidad-visible-a- traves-de-las-redes- sociales/?fbclid=IwAR09hK6AYsawhg1r813W dNmS9q3ZvOkHk5dsH- O8wQ6HUrHxS9fNhKGe6g

Vera, T. S. (2020). Perception of owners on the current laws of protection and animal ownership, applied in dogs treated in two veterinary clinics in the city of Guayaquil. Guayaquil: Repositorio de la Universidad Católica de Santiago de Guayaquil . http://repositorio.ucsg.edu.ec/bitstream/3317/14643/1/T-UCSG-PRE-TEC-CMV- 73.pdf

Annexes

Annexes 01: Consistency Matrix

Title: "Protection and welfare of domestic animals with respect to the policies implemented by the Provincial Municipality of San Martin, Tarapoto, 2020".

General problem formulation	Objectives	Hypothesis	Theoretical Framework	Variables	Specific variable	Means of registration	Unit of measurement	Methodological Framework
General Problem What is the relationship between the Protectionand wellbeingof domestic animals with respect to the policies implemented by the Provincial Municipality of San Martin, Tarapoto, 2020?	**General objective:** To determine the relationship between the protection and welfare of domestic animals with respect to the policies implemented by the Provincial Municipality of San Martin, Tarapoto, 2020. **Specific objectives** a) Identify the articles of the Law N° 30407 that prescribe on domestic animals and the knowledge of the population of the district of Tarapoto about them. b) Identify policies implemented by the Provincial Municipality of San Martin in the district of Tarapoto, with regard to domestic animals and the knowledge of the population of the district of Tarapoto about them. c) Identify compliance with the policies implemented by the	General Hypothesis: The relationship betwe en the protection and welfare of domestic animals and the policies implemente d by the Provincial Municipality of San Martin in the district of Tarapoto, 2020 is positive.	CHAPTER II THEORETICAL FRAMEWORK 2.1. Background to the investigation 2.1.1. International background 2.1.2. National background 2.1.3. Local background 2.2. Theoretical underpinnings 2.2.1. Protection and welfare of domestic animals 2.2.2.	Variable 1: Protection and welfare of domestic animals. Variable 2: Public Policies.	Cannot be measured	Surveys . Documentary analysis guide. Interview guide.	Nominal Nominal	Type and level of research Quantitative Correlational. Design Non-experimental Population Citizens of the district of Tarapoto Sample 300 citizens of the district of Tarapoto (Simple random) Data collection instruments The techniques to be used are the virtual survey (200 people), the physical survey (100 people), the documentary analysis guide (one for each regulation) and the interview guide. Data processing and analysis It will be processed through the Excel programme.

	Provincial Municipality of San Martin in the district of Tarapoto, with regard to domestic animals. d) Establish the relationship between the protection and welfare of domestic animals and the policies implemented by the Provincial Municipality of San Martin in the district of Tarapoto, 2020.		44			

Annexes 02: Photos surveying villagers in Tarapoto District

Figure 1. *Survey of the population in the Plaza de Armas of Tarapoto - Physical survey.*

Figure 2. *Survey of the population in Suchiche Park - Physical survey.*

Annex 03: Physical Research Instruments

physical research - Survey

Anexos 03: Instrumentos de investigación en físico – Encuesta

Survey addressed to the citizens of the district of Tarapoto.

Good morning we are graduates of the Faculty of Law and Political Science of the National University of San Martin, we are currently conducting research called: Protection and welfare of domestic animals with respect to the policies implemented by the Provincial Municipality of San Martin, Tarapoto, 2020. Therefore, we present a group of questions in order to answer our research through the answers obtained in this survey of 300 citizens of the district of Tarapoto.

Thank you in advance.

Name:

Date:

1   Do you know if there is any law that protects and provides for the welfare of domestic animals? (Objective A)

    Yes No

2   Are you aware of the existence of the Animal Protection and Welfare Law - Law N°30407 and its articles referring to domestic animals? (Objective A)

    Yes No

3   Do you know of any decree or ordinance that the authorities of the Provincial Municipality of San Martin have issued regarding Animal Protection and Welfare? (Objective B).

    Yes No

4   Do you think that the Provincial Municipality of San Martin is applying any policy in the district of Tarapoto in favour of the welfare and protection of domestic animals? (Objective C).

    Yes No

5   What level of protection do you think is given to domestic animals by the Provincial Municipality of San Martín (Objective C)?

    BadRegular GoodNone

6   Do you know if a municipal animal shelter has been created in the city of Tarapoto? (Objective D).

    Yes No

7   Do you know if awareness campaigns are being carried out in the city of Tarapoto regarding domestic animals? (Objective D).

    Yes No

8   Do you think that in the district of Tarapoto the Municipality is taking any positive action to help the situation of stray domestic animals? (Objective D).

Yes No
9   Did you know that domestic animal abuse can be reported (Objective D)?
    Yes No
10 Have you ever reported this type of abuse (Objective D)?
    Yes No

Annex 04: Validation sheet of the research instrument by expert judgement

NATIONAL UNIVERSITY OF SAN MARTÍN
FACULTY OF LAW AND POLITICAL SCIENCE
PROFESSIONAL SCHOOL OF LAW
Research: "Protection and welfare of domestic animals with respect to policies implemented by the
policies implemented by the Provincial Municipality of San Martin, Tarapoto,
2020".

1. **GENERAL DATA.**

1.1. **Surname and first name of the expert:** Jessica Elka Picón Pereda.

1.2. **Academic degree of the expert:** Lawyer.

1.3. **Surname and first name of researcher** : Alejandra Lizbeth Llontop Balarezo y

: Sharon Melissa Vilcarromero Ramírez

1.4. **Name of instrument** : Survey

1.5. **Author of the instrument** : Alejandra Lizbeth Llontop Balarezo and

: Sharon Melissa Vilcarromero Ramírez

II. CRITERIA FOR EVALUATING THE INSTRUMENT.

Indicators for assessing the instrument	Qualitative assessment	Deficient	Regular	Good	Very good	Excellent
	Quantitative assessment	**1**	**2**	**3**	**4**	**5**
Clarity	The items are formulated in appropriate language.				X	
Objectivity	The instrument allows for the collection of observable data or behaviour.					X
3. News	The instrument corresponds to the current state of knowledge.				X	
4. Organisation	There is a logical organisation of the items.					X
5.Sufficiency	The instrument assesses the quality and quantity dimensions of the					X
6. Intentionality	The instrument is adequate to achieve the objectives of the study.					X
7. Consistency	The instrument is based on scientific and theoretical aspects in line with the subject of the study.				X	
8. Coherence	The items relate to the variables dimensions and indicators.					X
9. Methodology	The instrument responds to the method, type, design and approach of the study.					X
10. Relevance	The items are suited to the type of research				X	
TOTAL						

III. EXPERT OPINION

The instrument was approved. It was found to be valid and applicable for the type of research.

Tarapoto, 11 January 2023.

EXPERTO

SURVEY APPLIED TO THE INHABITANTS OF THE DISTRICT OF TARAPOTO TO EVALUATE THE LEVEL OF KNOWLEDGE AND APPLICATION OF THE ANIMAL PROTECTION AND WELFARE LAW - LAW №30407.

Good day we are graduates of the Faculty of Law and Political Sciences of the National University of San Martin, at the moment we are carrying out the investigation denominated: Protection and well-being of domestic animals with respect to the policies implemented by the Provincial Municipality of San Martin, Tarapoto, 2020. Therefore, we present a group of questions in order to answer our research through the answers obtained in this survey of 300 citizens of the district of Tarapoto.

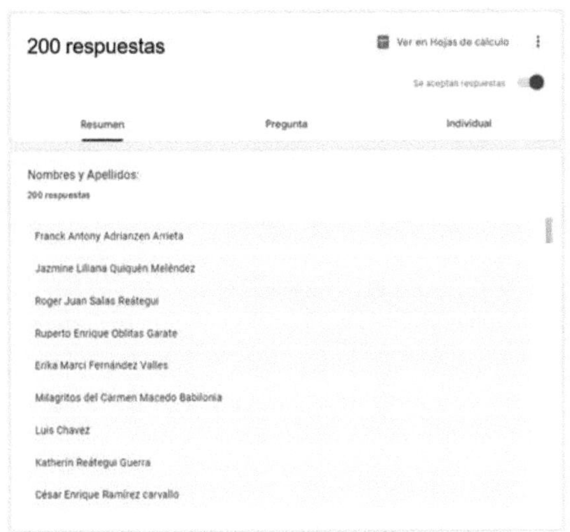

Thank you in advance.

Figure 1 and 2. In the following two photos we can see the survey carried out virtually with the inhabitants of the district of Tarapoto, which has the same content of questions as in Annex 3.

Annex 06: Photographs taken of domestic animals during the investigation

Figura 1. *In the following photo we can see a canine wandering along Av. N°6 at the height of the airport in the district of Tarapoto, which was found in very bad health conditions, with difficulty to walk, to eat and without teeth.*

Figura 2. *In the following photo we can see two dogs wandering in the vicinity of Jr. Manco Inca cdra. N°1, which roam the streets without a collar and unaccompanied by any person.*

Figura 3. *In the following photo we can see two dogs on a corner in Jr. Leoncio Prado cdra. N°13, which are sleeping in the sand because of the cold of the night, they do not have collars or anything that identifies them.*

Figura 4. *In the following photo we can see three dogs in Jr. Miguel Grau cdra. N°1, which are looking for food in the rubbish and do not have a collar that identifies them.*

Figure 5. *Jr. Miguel Grau cdra. N°1.*

Figure 6. *In the following photo we can see four dogs roaming freely at night without a collar, harness or muzzle in Jr. Ramon Castilla cdra. N°2.*

Figure 7. *Jr. Ramon Castilla cdra. N°2.*

Figure 8. *Jr. Miguel Grau cdra. N°1.*

Figure 9. In the following photo we can see two dogs without collars that identify them, sleeping at night in the middle of the street where motorised vehicles circulate in Jr. Los Proceres cdra. N°3.

Annex 07: Request for Access to Public Information

Fecha: 10/04/2023

hora: 11:11:28

T-1680-2023-MP-SG-MPSM.JRRP
Trámite Libre
N° Documento . **Solicitud**
Asunto : **Acceso A La Informacion Publica**
Solicitante: - Vilcarromero Ramirez Sharon Meli
Usuario: José Rodvin Reátegui Pezo

Destino
Oficina: Secretaria General
Cargo: Asistente
Nombre: 07617498 - Alfredo Del Aguila Lina
Fecha: ..../...../..........   Hora: ....

Firma. ....

Observacion: ....

SOLICITO: <u>ACCESO A LA INFORMACIÓN PÚBLICA.</u>

SEÑORA LUZ ADELINA SAAVEDRA RENGIFO
Gerente De Recursos Naturales Y Gestión Ambiental

SHARON MELISSA VILCARROMERO RAMÍREZ, identificada con **DNI N° 71721627**, con correo electrónico shavilram@gmail.com, celular N° 945527335, con domicilio en **Jr. Los Pinos N° 653 - Tarapoto**, ante usted con el debido respeto me presento y digo:

Que, en mi calidad de pobladora de la ciudad de Tarapoto, en virtud del Texto Único Ordenado de la Ley N° 27806, Ley de Transparencia y Acceso a la Información Pública, aprobado por Decreto Supremo N° 043-2003-PCM, Reglamento de la Ley de Transparencia y Acceso a la Información Pública, DECRETO SUPREMO N° 072-2003-PCM, publicado el 07-08-2003, *solicito ante su despacho, me facilite copia simple de todas las ordenanzas municipales y demás normativa emitida por la Municipalidad Provincial de San Martín, respecto a la tenencia, cuidado, protección y bienestar de animales domésticos hasta el año 2020.*

Pongo de conocimiento también que, dicha información será utilizada estrictamente para fines académicos y autorizo la remisión de la misma vía correo electrónico: shavilram@gmail.com o de lo contrario a la dirección física indicada en el exordio.

<u>POR LO EXPUESTO:</u>
Ruego a usted acceda a mi solicitud por ser de justicia, agradezco de antemano su atención.

Tarapoto, 10 de abril del 2023

____
SHARON MELISSA VILCARROMERO RAMÍREZ
DNI N° 71721627

MUNICIPALIDAD PROVINCIAL
DE SAN MARTÍN

"Año de la Unidad, la paz y el desarrollo"

CARTA N° 024-2023-SGDEL-GDE/MPSM.

Señor (a) :
SHARON MELISSA VILCARROMERO RAMIREZ

REFERENCIA   : MEMORANDO N° 0179-2023-GDE-MPSM

EXPEDIENTE   : T- 1680-2023-MP-SG-MPSM.JRRP /
                    MEMORANDO N° 173-2023-SG-MPSM

Es grato dirigirme a usted, para saludarle cordialmente a nombre de la Sub Gerencia de Desarrollo Económico Local de la Municipalidad Provincial de San Martín y en atención al documento de la referencia, en cumplimiento de la Ley N° 27806, Ley de Transparencia y Acceso a la información Pública.

La Municipalidad Provincial de San Martín cuenta con la Ordenanza Municipal N°014 - 2020 – MPSM; REGLAMENTO QUE REGULA EL REGIMEN DE TENENCIA Y REGISTRO DE CANES EN EL DISTRITO DE TARAPOTO; siendo está toda la información concerniente a la tenencia, cuidado, protección y bienestar de animales domésticos.

Sin otro particular me suscribo de Usted.

Tarapoto; 13 de abril del 2023

Atentamente,

Sharon Melissa Vilcarromero R.
71721627

MASH/SGDE
C.c.
Archivo.

M.V. MARCO A. SÁNCHEZ HUARIPATA
SUB GERENTE DE DESARROLLO ECONÓMICO LOCAL
MUNICIPALIDAD PROVINCIAL DE SAN MARTÍN
TARAPOTO

Jr. Gregorio Delgado N° 260, Tarapoto
www.mpsm.gob.pe
(042) 522151 / mpsm@mpsm.gob.pe

Annex 09: Analysis Guide
Documentary 01- LAW N° 30407.

Name of document: Law No. 30407: "Law on Animal Protection and Welfare".

Objective: To analyse the regulations, with the aim of contrasting what is established in the regulation with the results of the survey and expert judgement, in order to be able to conclude whether what is regulated is effectively applied in our locality.

Criteria for analysis:

1.   General reading of the Law.

The following is a detailed reading of each of the articles prescribed in Law No. 30407, Law on the Protection and Welfare of Animals. From the general reading it is clear that this law covers the protection and welfare of both domestic and wild animals.

2.   Determine the articles prescribing the protection and welfare of domestic animals.

Having carefully read the law, we determined that the articles prescribing the protection and welfare of domestic animals are as follows:

PRELIMINARY TITLE

-   Article 1. Principles.

CHAPTER I

GENERAL PROVISIONS

-   Article 2. Purpose.

-   Article 3. Object of the Law.

-   Article 4. Definitions.

CHAPTER II

DUTIES OF INDIVIDUALS AND THE STATE

-   Article 5. Duties of persons:

-   Article 6. Complaint for non-compliance with the law

-   Article 7. Duties of the State:

-   Article 8. Temporary Shelters:

CHAPTER III

GOVERNING BODY AND IMPLEMENTING AND SUPPORTING BODIES

-   Article 9. Governing body and inter-sectoral coordination.

-   Responsibilities of the authorities and institutions involved.

-   Article 11. Animal protection and welfare committees.

CHAPTER IV.

ANIMAL PROTECTION AND WELFARE ASSOCIATIONS

- Article 12. Animal protection and welfare associations.

- Article 13. Registration and accreditation of animal protection and welfare associations.

CHAPTER V.

KEEPING, PROTECTION AND MANAGEMENT OF ANIMALS.

- Article 14. Animals as sentient beings.

- Article 15. Professional staff.

- Article 21. Protection and welfare measures for companion animals or pets.

CHAPTER VI.

PROHIBITIONS

- Article 22. General prohibitions.

- Article 25. Prohibitions and exemptions for the use of animals in experiments, research and teaching. Article 27. Prohibition of harm to companion animals.

CHAPTER VII.

EUTHANASIA OF DOMESTIC COMPANION ANIMALS AND WILD ANIMALS KEPT IN CAPTIVITY

- Article 28. Consent to and performance of euthanasia.

- Article 29. Methods of euthanasia

CAPÍTULO VIII

INFRINGEMENTS AND PENALTIES

- Article 30. Infringements and penalties.

- Article 31. Interim measures

- Article 32. Persons responsible for the infringement

- Article 33. Administrative, civil or criminal liability

- Article 34. Financing

3. Determine the substantive content of each of these articles.

PRELIMINARY TITLE

- Article 1. Principles: This article prescribes the guidelines on which the law is based, which is why it also covers the protection and welfare of domestic animals.

CHAPTER I

GENERAL PROVISIONS

This chapter prescribes the general provisions on which this law is based and therefore also covers domestic animals:

- Article 2. Purpose: Seeks to ensure the welfare and protection of vertebrate animals, including domestic animals, taking into account animal life, animal health and public health.

- Article 3. Object of the Law: To safeguard the health and life of animals, including domestic animals, giving zero tolerance to cruelty and mistreatment, promoting respect for the life and welfare of animals through education. It also seeks to care for the welfare of animals by avoiding accidents to their populations and also preventing diseases that, when transmitted by them, cause damage to human health. It seeks to promote the participation of public and private entities together with the entire population, subject to the constitution.

- Article 4. Definitions: Here we find the definitions that the law takes into account when referring to animal welfare, companion animals and domesticated species.

CHAPTER II

DUTIES OF INDIVIDUALS AND THE STATE

Article 5. Duties of persons: It establishes that every person has the duty to seek the welfare of any animal species, including domestic animals; the ownership and acquisition of an animal is the responsibility of a capable person; the owner must attend to the animal's fundamental needs: provide an adequate habitat, with minimum sanitary conditions, protection from disease and pain, veterinary care and vaccinations.

- Article 6. Reporting non-compliance with the law. It empowers any person to denounce any infraction of the present law, and authorities such as the Public Prosecutor's Office, PNP and local governments are obliged to attend to these complaints, guaranteeing the application of the law.

- Article 7. Duties of the State: It has the duty to guarantee the protection, health and life of domestic animals.

- Temporary shelters: It is the responsibility of local governments to promote the creation and operation of temporary shelters for abandoned domestic animals. The veterinary medical association is also empowered to support the creation of minimum technical standards for this work.

CHAPTER III

GOVERNING BODY AND IMPLEMENTING AND SUPPORTING BODIES

- Article 9. Governing body and inter-sectoral co-ordination: The governing body is the Ministry of Agriculture, being in charge of channelling the input of the Ministry of Health (when human health is concerned), the Ministry of Environment (biodiversity) and the Ministry of Education (education with regard to animal protection and welfare).

- Responsibilities of the authorities and institutions involved: A. Ministry of Education: Develops the environmental issue, promoting the protection and defence of biological diversity; B. The executive branch, national, regional and local governments: Oversee that the law is applied; C. Local governments: Supervise compliance with the law when granting licences and authorisations related to the possession, trade, transport and veterinary care of animals. D. The institutions involved, whether public or private, in case of emergency, provide relief to animal protection and welfare associations so that they can rescue, protect and assist the animals.

- Article 11. Animal protection and welfare committees: It is the responsibility of the regional governments to establish a regional animal protection and welfare committee to be headed by the regional governor, together with the provincial mayors or their representative, a representative of the animal protection and welfare associations and a representative of the professional associations of biologists, doctors and veterinary doctors of Peru. Its functions are:

a.   Coordinate and liaise with relevant sectors.

b.   Issue technical reports on matters within its competence.

c.   Propose by-laws for the enforcement of animal protection and welfare measures in its jurisdiction.

d.   Collect, systematise and make information on the non-responsible keeping of pets and experimental animals available to the competent bodies for the necessary actions.

e.   Establish a register of Research Ethics and Animal Welfare Committees of user centres.

f.   Issue annual reports or balance sheets on its activities, which are submitted to the relevant sectors.

g.   implement provincial animal protection and welfare committees within one year of the entry into force of this law.

CHAPTER IV.

ANIMAL PROTECTION AND WELFARE ASSOCIATIONS

-   Article 12. Animal protection and welfare associations: These are non-profit legal entities that are legally constituted, whose objective is to protect and defend animals.

-   Article 13. Registration and accreditation of animal protection and welfare associations: The procedures, obligations and requirements for local, regional and national registration and accreditation of animal protection associations are established by the Ministry of Agriculture.

CHAPTER V.

KEEPING, PROTECTION AND MANAGEMENT OF ANIMALS.

-   Animals as sentient beings: Domestic and wild animals in captivity are considered as sentient animals by law.

-   Article 15. Professional staff: Any person, whether natural or legal, carrying out research or animal protection and welfare activities, must have professionals for each species.

-   Article 19. Centres that use animals for experimentation, research and teaching: Animal experimentation and research will only be carried out in higher education and specialised centres that have animal welfare ethics committees with measures oriented towards good bioethical and biosecurity management practices, according to each species, taking into account what has been established by the Ministry of Agriculture.

-   Article 20. National ethics committee for animal welfare: The committee is made up of six members: a representative of the national forestry and wildlife authority of the ministry of agriculture and irrigation, the ministry of the environment, the national agricultural health authority, the national council for science, technology and technological innovation (CONCYTEC), the Peruvian veterinary medical association and the Peruvian association of biologists.

This committee is in charge of qualifying the guidelines used by the ethics committees of each centre to set guidelines based on international criteria.

- Article 21. Protection and welfare measures for companion animals or pets: Owners, managers, those responsible for commercial establishments, breeding centres for captive breeding, security services, training, the national police, the fire brigade, the forces, municipalities or any public or private entity and any person who has domestic or wild animals must comply with the animal protection and welfare measures established by the Ministry of the Environment together with the Ministry of Health.

These protection and welfare measures are based on good practices on trade, breeding, adoption, quarantine, transport and keeping approved by the competent sectors.

CHAPTER VI.

PROHIBITIONS

- Article 22. General prohibitions: All activities which may endanger the protection and welfare of animals such as:

A. Leaving animals on public roads creates abuse and endangers public health. The regional and local governments are responsible for controlling and imposing sanctions for acts of animal abandonment.

B. Using animals by conditioning them to behave in a certain unnatural way to entertain people in a public or private environment.

C. Keeping, hunting, trapping, breeding, buying and selling animals that are not classified as food animals.

D. Domestic animal clashes in public or private environments.

- Article 25. Prohibitions and exemptions for the use of animals in experiments, research and teaching: It is prohibited:

A. All experimentation and research on live animals which causes unnecessary pain, laceration or death, unless it is essential for scientific research and the results of the experiments cannot be achieved by other means, cannot be replaced by other means and are essential for:

1 To control, prevent and diagnose diseases that harm humans or animals.

2 Assessment, determination, regulation or modification of physiological conditions in humans and animals.

3 Protecting the environment and preserving biodiversity.

4 Investigation of animal production parameters.

5 Medico-legal enquiry.

B. Experimenting with any species of animal in teaching or research activities in any educational institution regardless of grade or category.

- Article 27. Prohibition of attacks on companion animals: Any behaviour that endangers the protection and welfare of companion animals is prohibited, such as:

- A. Surgical amputation or surgery that is unnecessary or interferes with the animal's ability to express natural behaviour, only surgery that is clinically indicated is acceptable.

- B. Training, promotion and organisation of animal fights.
- C. Keeping and occupying pet animals for human consumption.
- D. Commercial use of processed pet animal products and by-products.
- E. Uncontrolled commercial exploitation affecting the welfare of pets.
- F. The number of animals kept exceeds the number that the owner is permitted to keep under this Act.

CHAPTER VII.

EUTHANASIA OF DOMESTIC COMPANION ANIMALS AND WILD ANIMALS KEPT IN CAPTIVITY

- Article 28. Consent and execution of euthanasia: The killing of domestic animals can only be carried out at the suggestion of a veterinary doctor or zootechnician, duly registered and authorised, and with the permission of the owner of the animal in a document in his or her own handwriting.

- Euthanasia methods: For situations that represent a danger to public health, the Ministry of Health is in charge of defining inspection procedures in accordance with the aforementioned law, where only procedures that do not cause pain or suffering and under the supervision of the veterinary doctor will be consented to.

CHAPTER VIII

INFRINGEMENTS AND PENALTIES

- Article 30. Infringements and penalties:

30.1. Failure to comply with the obligations and prohibitions set out in Article 5 and Chapter VI "Prohibitions" of this law constitutes an administrative offence.

30.2. Sanctions imposed by competent Ministries are considered in the present regulation, in which regional and local governments have the power to sanction within their material and territorial attributions, as well as to ensure compliance with the obligations in accordance with the applicable legislation.

30.3. The administrative sanctions are:

A. A fine of not less than one and not more than fifty UIT.

B. Suspension of both experiments and research if they do not comply with the above-mentioned law.

C. Partial or total, temporary or permanent closure of the offending centres or institutions.

D. Confiscation of an object, instrument or device used in the commission of an offence.

E. Suspend or cancel a licence, operating permit, concession or any other permit, respectively.

Sanctions are applied in accordance with the General Administrative Procedure Law No. 27444, the principle of reasonableness established in Article 230 of the same law.

- Article 31. Provisional measures: The provisions of paragraph 30.3 of Article 30, i.e.

letters B, C, D and E, are introduced provisionally before the initiation of the administrative sanctioning procedure, and may also be imposed during the development of the procedures described above.

\- Article 32. Persons responsible for the offence: Anyone who, by his action or omission, contributes to the commission of an act in violation of this law, shall be responsible for the reparation of the offence. The owner or keeper of one or more animals, the person responsible or the owner of the sitting, enclosure or property and the owner of the transport company or the owner of the vehicle, or the driver where the offence occurred, depending on the situation.

\- Article 33. Administrative, civil or criminal liability: The administrative obligation is independent of any civil or criminal liability that may arise from the infringement.

\- Article 34. Financing: The activities required for the implementation of this law are carried out with the budgets of the institutions in the corresponding specifications, without requesting additional funds from the public treasury.

4. From the above, which articles prescribe the role or intervention of local governments in animal protection and welfare?

The articles are as follows:

\- Article 6. Complaints of non-compliance with the law: The law establishes that local governments must address complaints and intervene to guarantee the application of the law.

\- Article 8. Temporary Shelters: The law establishes that local governments must promote the creation and also the operation of temporary shelters for abandoned domestic animals.

\- Responsibilities of the authorities and institutions involved: The law establishes that local governments: Oversee the application of the law and supervise compliance with it, when granting licences and authorisations related to animal ownership, commercialisation, transport and veterinary care.

\- Article 11. Animal protection and welfare committees: The law establishes that it is the responsibility of the regional governments to establish a regional animal protection and welfare committee, which must be headed by the regional governor and made up of the provincial mayors or their representative, a representative of the animal protection and welfare associations and a representative of the professional associations of biologists, doctors and veterinary doctors of Peru. Its functions are:

a. Coordinate and liaise with relevant sectors.

b. Issue technical reports on matters within its competence.

c. Propose by-laws for the enforcement of animal protection and welfare measures in its jurisdiction.

d. Collect, systematise and make information on non-responsible pet and experimental animal ownership available to the competent bodies for the necessary actions.

e. Establish a register of Research Ethics and Animal Welfare Committees of user centres.

f.   Issue annual reports or balance sheets on its activities, which are submitted to the relevant sectors.

g.   Implement provincial animal protection and welfare committees within one year of the entry into force of this law.

-   Article 22. General prohibitions. This article establishes that local governments have the power to provide the necessary mechanisms to stop the abandonment of animals and to impose the corresponding sanctions.

-   Article 30. Infringements and sanctions: The law establishes in paragraph 2 of this article that regional governments and local governments have sanctioning powers according to their material and territorial competences; in turn, they carry out the coercive enforcement of the obligations derived from the law.

5. Make a conclusion of what is enshrined in the regulations regarding domestic animals and the roles of local governments.

1)   Law N° 30407: "Animal Protection and Welfare Law", has 26 articles that are applicable to the protection and welfare of domestic animals, 6 of these articles establish the role and intervention of local governments in the application of this law.

2)   Law No. 30407, with regard to domestic animals, regulates the duties of people and the state, the governing body and executing and supporting bodies, animal protection and welfare associations, animal ownership, protection and management, euthanasia of domestic pets, and finally infringements and sanctions in the event of non-compliance with the law.

3)   On the intervention and role of local and regional governments, we have that the law establishes that:

a)   They have a duty to address complaints and intervene to ensure law enforcement,

b)   They should encourage the creation and also the operation of temporary shelters for abandoned pets.

c)   To monitor the application of the law and supervise compliance with it, when granting licences and authorisations related to the keeping, marketing, transport and veterinary care of animals.

d)   They should be part of the animal protection and welfare committee.

e)   They are empowered to provide the necessary mechanisms to control the abandonment of animals and to impose the corresponding sanctions.

f)   Finally, it establishes that regional and local governments have the power to impose penalties within the scope of their material and territorial competences; they also carry out the coercive enforcement of the obligations derived from the law.

Objective: To analyse the articles of the regulation, with the aim of contrasting what is established in the regulation with the results of the survey, in order to be able to conclude if what is regulated is effectively applied in our locality.

Criteria for analysis:

1. General reading of the municipal ordinance.

We proceed to read carefully each of the articles of the Municipal Ordinance N.° 014-2020-A/MPSM. This ordinance approves the regulation on the regime of dog ownership and registration in the district of Tarapoto and consists of 36 articles.

2. Determine the articles prescribing the protection and welfare of domestic animals.

Taking into account the issue of regulations, regulations on the regime of dog ownership and registration in the district of Tarapoto, we consider that the 36 articles involve the protection and welfare of domestic animals.

3. Determine the substantive content of each of these articles.

REGULATIONS GOVERNING THE SYSTEM OF DOG OWNERSHIP AND REGISTRATION IN THE DISTRICT OF TARAPOTO.

TITLE I

GENERAL PROVISIONS SUBJECT MATTER AND SCOPE OF APPLICATION

Article 1.- This regulation governs the registration and ownership of dogs in the district of Tarapoto, with the aim of safeguarding the integrity of citizens and pets.

Article 2.- The scope covers the breeding, keeping, sanitary control and protection of dogs in the district of Tarapoto, in order to ensure that they develop in a suitable environment.

TITLE II

BREEDING AND KEEPING OF DOGS

Article 3.- A dog is considered a domestic animal of company for human beings that easily responds to practices, any natural or juridical person can raise them or possess them if it has a suitable environment so that its possession does not alter in any way the tranquillity and the well-being of the citizenship. For this reason, ownership will depend on the conditions for the welfare and health of the dog.

Article 4.- In the case of buildings, the neighbourhood councils shall decide on the guidelines for the breeding or keeping of dogs and shall report their decisions to the municipal authorities. Agreements shall not contravene the provisions of law 27 59 6, if there is no neighbourhood council, decisions shall be authorised by the signature of the majority of the neighbours.

Article 5.- According to Ministerial Resolution N° 1776 - 2002-SA/DM, the following breeds are considered dangerous breeds due to their aggressive fighting and attacking characteristics as well as their genetic load: Dogo Argentino, File Brasileiro, Bull Mastiff, Japanese Tosa, Rottweiler, Doberman and the hybrid American Pitbull Terrier. Any dog that has a history of being aggressive towards people and other animals or has been trained for fighting is considered dangerous.

Article 6.- All owners, keepers, guardians of dogs that do not live in the neighbourhood of the jurisdiction of the district of Tarapoto must be responsible for collecting all organic waste left by the dog and must also deposit it in rubbish bins or containers designed for this purpose. In case of non-compliance, a monetary fine may be imposed, complaints may be made and all that is required by law.

Article 7.- In the district of Tarapoto it is forbidden to implement informal breeding centres and/or commercialisation of meat. If breeding centres exist, they must be located in suitable environments run by suitable people and with the help of a registered and authorised veterinary doctor.

Article 8.- Each owner of a dog must be responsible for having a space within his or her home to bury the dog upon its death. If they do not have this space, they can locate one outside the city.

TITLE III.

OF THE REQUIREMENTS FOR OWNING OR KEEPING DOGS

Article 9.- The requirements to breed and own a dog in the district of Tarapoto are: to be over 18 years of age, to have a fixed address in the district, to have a physical environment with hygienic conditions for adequate breeding and ownership, to have the dog vaccinated and to have all the security mechanisms for walking it in public spaces, and to have a sworn declaration stating that it has not been sanctioned with respect to the provisions of Law No. 27596 and its regulations during the last two years from the date of application.

Article 10.- The requirements for owning a dangerous dog are: to have the capacity to enjoy and to be of legal age, to have a psychological attitude certified by a registered psychologist and not to have been sanctioned in accordance with the law in the three years prior to owning or acquiring the dog.

TITLE IV

OF THE CLASSIFICATION PROCEDURE, OBLIGATIONS AND DUTIES OF THE OWNER OR KEEPER OF DOGS

Article 11.- The sub-management of local economic development may classify a dog as dangerous or potentially dangerous:

- For his aggressive behaviour at the request of no less than 15 neighbours identified with ID and indicating that they are from an address close to the dog.
- The owner or person in possession of the dog must be notified so that he/she can give his/her defence and if this defence is declared inadmissible, the dog will be considered as dangerous or potentially dangerous, this decision being irrevocable and obliging the owner to take the actions set out in the regulations.

A dog is considered potentially dangerous when:

- Without being provoked, he/she should not be threatening to neighbours in a public space.
- Any person approaching its owner's property reacts by attacking in a terrifying manner in the presence of its owner.
- They have been trained for attack or defence.
- Have killed or assaulted domestic animals belonging to their owner or to another person.

Article 5 of the regulation should also be taken into account.

A dog is considered dangerous if it:

- Regardless of its breed, it is aggressive or bites another animal or human being, except those whose behaviour is due to panic attacks or stress generated by natural phenomena or celebrations, mishandling or mistreatment.
- has injured any person causing permanent disability.
- Has killed another dog without being provoked to do so.
- Been trained to fight.

Article 12.- The person who owns a dog is obliged to keep it in hygienic conditions and to apply its preventive treatments. It is forbidden:

a) Training that reinforces aggression.

b) The entry of dogs considered potentially dangerous into public premises where there is mass attendance.

c) Entry to children's play areas.

Article 13.- The owners or keepers of dogs are responsible when:

a) If your dog causes serious injury to a person, the owner shall cover the full cost of medication, hospitalisation and surgery until the person has fully recovered, without prejudice to compensation for damages. This provision shall not be compulsory when the dog is acting in self-defence of third parties or any private property.

b) The dog causes injury to another animal. Its owner must cover the full cost of the recovery of the attacked animal, if it dies, the owner of the attacking dog must assume criminal prosecution and civil compensation in accordance with the law. This does not apply if acting in defence of self-defence, third parties and private property.

Article 14.- Other duties of the owners and keepers of animals apart from those established by law:

a)   Register and identify your dogs that you own or have custody of.

b)   Obtain the licence.

c)   Walking dogs in public places on leashes. Potentially dangerous dogs must additionally be muzzled. These dogs must be led by persons who are physically and mentally capable of exercising control over them.

d)   Keep dogs in a safe manner avoiding harm to the third party.

e)   Manage the licensing of your pets' offspring.

TITLE V

TRAINING, MARKETING AND TRANSFER OF DOGS.

Article 15.- The following shall be considered when training a dog:

1)   It must be carried out in an authorised and qualified centre for this purpose, offering the necessary security so as not to endanger the integrity of the persons involved.

2)   These training, care and trade centres shall:

a)   Have a municipal operating licence.

b)   Have health authorisation issued by Essalud.

c)   Have a positive report from a State Cynological Organisation.

d)   Have trained personnel and protective equipment.

e)   To have hygienic facilities and environments that make it easier for dogs to move around and to have water and food tanks.

f)   Avoiding noise that disturbs neighbours.

g)   Permanent and proper disposal of dog faeces.

h)   Report any zoonosis to the health authorities.

1) If the establishment does not comply with the provisions of the municipality, it is empowered to close it down and close it definitively, without prejudice to placing the dogs in shelters as provided by the authority.

2) It is forbidden for training centres to direct their training towards strengthening the dog's aggressiveness. They are also prohibited from organising or conducting dog fights in private or public places.

Article 16.- In order to trade or transfer a dog, the following must be considered:

a)   Natural or legal persons may sell dogs in authorised places, taking into account the regulations; therefore, the seller has the obligation to indicate to the buyer everything concerning the breed of the dog, especially those considered potentially dangerous.

b)   In order to obtain municipal authorisation, the marketing establishment must certify that it will be run by a registered veterinary doctor who is authorised to practise his profession. This professional must also have a psychological certificate attesting to his or her aptitude.

Article 17.- If potentially dangerous dogs are to be marketed, the following shall additionally be complied with:

a)   Legal persons must designate a natural person to be responsible for the protection and care of the dog.

b) Natural persons must comply with the provisions of Articles 4 and 5 of Law No. 27596.

Article 18.- Those who trade, sell or donate dogs shall provide the buyer with adequate information on the character of the animal and basic guidelines for proper breeding.

TITLE VI

IDENTIFICATION, REGISTRATION AND LICENSING OF DOGS

Article 19.- The owners of animals considered potentially dangerous, within a period of no more than three months from their birth or one month from their purchase, are obliged to register them in the municipal register of potentially dangerous animals, indicating the details of the owner of the animal, the place where it will live and specifying whether it is destined to live with people or whether its purpose is to keep or protect.

Article 20.- Stray and abandoned animals:

a) A stray animal is an animal that moves within the jurisdiction of the city without identification and without the company of humans.

b) An abandoned animal is an animal which, even if it has identification, is wandering around unaccompanied by humans, nor has the police been informed of its loss by the owner or an authorised person.

c) Due to inter-institutional agreements with public or private animal welfare institutions signed by the Provincial Municipality of San Martin, stray and abandoned animals will be moved to shelters, the owners have 30 days to pick up their animals, as indicated in Law N°27596 of the Legal Regime of Dogs and its regulations, in addition, the owner or the person with authorisation who picks up the animal must present the documents that identify the animal to the respective shelter according to the regulations of the D. S. N°006-2002-SA and also the anti-rabies vaccination card.S. N°006-2002-SA and also the anti-rabies vaccination card, otherwise, in the absence of any documentation, the owner of the animal must pay the corresponding fee for the vaccination of the animal, also, whatever the situation, a sanctioning file will be processed against the owner of the dog in accordance with the regulations.

d) The Municipal Zoonosis Prevention Centre is in charge of disposing of the animals that were transferred to the shelter on duty and were not claimed by their owners, this will ensure the future of the animal according to Law 25796 which establishes the legal regime of dogs.

Article 21.- Guard dogs.

a) Guard dogs should be supervised by owners or authorised persons in establishments where they cannot cause any harm and where there should be a warning that a guard dog is present.

b) In establishments that are not enclosed, it is mandatory to have a shelter or an area where the animal is protected from the high temperatures.

c) Animals should be free to move, not kept tethered, but if necessary they should have some freedom to move.

Article 22.- Guide dogs.

a)   A guide dog is considered to be an animal that has been accredited for training in national or foreign centres, specifically for companionship, transport, support of the visually impaired or motor handicapped.

b)   Guide dogs can travel on any means of public - urban transport, as well as free entry to establishments, sites and public shows.

c)   The owner is responsible for the good behaviour of the animal in the event of any damage caused by the animal to a third party.

Article 23.- In order for the Municipality to receive a permit to keep or possess a dog, the following conditions must be fulfilled:

a)   Application addressed to the Deputy Manager of Local Economic Development.

b)   Copy of DNI of the owner or possessor.

c)   Health certificate or vaccination card of the animal issued by a registered and authorised veterinarian, this document must contain the details of the owner or keeper of the animal with his home address, breed, peculiarities or markings of the animal for quick identification, vaccinations received and veterinary history.

d)   An affidavit in accordance with Article 10 of the Regulation.

e)   02 Full body colour photos of the animal.

f)   Payment slip for registration or renewal of authorisation.

g)   Positive psychological evaluation of the owner or keeper in the case of dogs in the category of being very dangerous or potentially dangerous.

h)   If an owner transfers his dog to a third party he must do so with all the respective documentation of the animal.

Article 24.- In the case of registration of the "American Pitbull Terrier" and of dogs considered to be very dangerous, the owners must take out civil liability insurance in case the animal causes damage to a third party, and must also provide the Municipality with all the safety and protective clothing for the animal for its transport in public places.

Article 25.- Once the registration of the animal has been accepted, the municipal authority will provide the owner or keeper with an identification card and a metal tag containing a registration code, and it will therefore be compulsory for the animal to wear it on its collar.

Article 26.- Once the animal has been registered, in the event that the owner or keeper changes his address, transfers his dog to a third party, loses the animal or the animal dies, this must be communicated to the municipal authority.

Article 27.- Animal health certificates are renewable on an annual basis, and must be presented to the respective authority at the time they are requested and also when the owner cedes his dog to a third party.

TITLE VII

OF THE DETENTION OF DOGS

Article 28.- The Municipality may impose administrative sanctions within its jurisdiction on anyone who disregards these regulations in accordance with the RASA and CISA approved by ordinance N°019-2019-a/MPSM, as well as relocate the animals for a maximum period of 30 days in shelters authorised by the Municipality, the cost of this service provided to the animal shall be borne by the owner or keeper.

Article 29.- The municipal authority, in co-ordination with animal welfare institutions, shall be responsible for reintegrating into society animals that are abandoned in public places and whose owner or keeper cannot be identified; if this is not the case, it shall freely dispose of the animal.

Article 30.- In the confirmed case that a dog bites or causes injury to a person or animal, independently of whether or not it has an owner or possessor, the animal will remain under anti-rabies observation for a period of 10 consecutive days from the occurrence of the event, in the environment stipulated by the municipal authority, at the end of the observation process the veterinary doctor will issue a report on the progress of the animal's health and the affected persons will be notified for the restitution of the animal, and if necessary, the corresponding measures will be adopted.

Article 31.- If an animal causes serious physical harm or death to people or animals, it will have to be isolated and then euthanized. Serious physical harm is considered to be that which requires medical or veterinary attention and in which the wounded animal needs physical rest for a period of more than 15 days. Excluded from being deprived of their life are the animals that acted in aid of their owner, defence of themselves and their offspring, according to paragraph 11.3 of art. 11° of D.S. 006-2002-SA and paragraph 15.3 of the article, in addition this is in case the fact has been proven, according to the established in article 15° of the Law 27596.

TITLE VIII

OF INFRINGEMENTS

Article 32.- Minor infringements may be fined up to 0.5 UIT:

a)   Failure to register animals in the municipal register.

b)   Disobey all the provisions of Article 4 of Law N°27596, which states what it means to be the owner or keeper of a potentially dangerous dog.

Article 33.- The following serious infringements shall be punished with a fine of up to 1 UIT:

a)   Transporting an animal in public places without carrying its identification, without a muzzle or leash, or in the situation where such implements are not sufficient to restrain the animal according to its characteristics, as well as the person transporting the animal is not competent to do so, all this in the category of potentially dangerous dogs.

b)   Not being authorised to keep dogs.

c)   Failure to report each year in the municipal register the animal health certificate.

d)   Driving animals not complying with the provisions of Article 9 of Law No. 27596, in such a situation, apart from the keeper or owner, the driver is also fined.

e)   Failure to comply with the provisions of Article 3 (c) of Law N°27596, where, in addition to the owner or possessor, the natural or legal person responsible for the protection of the spectacle is fined.

Article 34.- The following very serious infringements will be sanctioned with a fine of up to 2 UIT:

a)   Cooperating, arranging, encouraging or publicising dog fighting.

b)   Educating dogs to cope with, increase or strengthen their aggressive temperament.

c)   To abandon dogs belonging to the category of potentially dangerous dogs.

Article 35.- As long as the fine is not cancelled and the infraction is not solved, the animal will be retained by the municipal authority, who will charge a daily fee for the care of the animal for the duration of its stay.

Article 36.- The severity of the sanction shall depend on the risk caused to society, the repetition of committing such infringements, as well as the monetary gain obtained from the infringement.

4. As a result of the above, make a general conclusion of what is set out in the regulation.

Municipal Ordinance N° 014-2020- MPSM, establishes the regulations for dog ownership and registration in the district of Tarapoto. From all the articles detailed above it can be deduced that:

•   The regulation consists of 36 articles, which only cover the keeping and registration of dogs, and does not cover any other domestic animals.

•   Within the dog ownership and registration regulations, we can identify that the authority tries to provide welfare and protection to dogs, indicating how they should be cared for and kept; it also establishes sanctions for owners and/or keepers in case of non-compliance.

•   It indicates that the training must be carried out in an authorised centre and must not be aimed at strengthening the dog's aggressiveness, and it is strictly forbidden for these centres to organise dog fights.

•   Dogs must be marketed in authorised places and the seller must provide the buyer with all the information concerning the breed of the dog.

•   The registration of potentially dangerous dogs must be carried out within a maximum period of three months from birth.

•   It establishes the signing of inter-institutional agreements with public or private animal welfare institutions to shelter abandoned dogs.

•   Guide dogs may travel on all means of transport, as well as have free admission to establishments, places and public performances.

- It provides that, in the case of adoption of very dangerous or potentially dangerous dogs, owners must have a positive psychological evaluation certificate.
- In the situation where a dog bites or causes injury to a person or animal, the dog shall be kept under observation and in the event of serious injury or death, the dog shall be put down, except for a dog that acted in defence of its owner, itself or its offspring.
- Minor offences are fined up to 0.5 UIT, serious offences up to 1 UIT and very serious offences up to 2 UIT. The severity of the sanction depends on the danger caused to society, whether the offence is repeated and the profit generated by committing the offence.

Annex 11: Invitation 01

Interview with experts in the field of environmental law.
04 July 2023
Tarapoto - Peru
Dear veterinary doctor Marco Antonio Sánchez Huaripata:
First of all, a cordial greeting to you. Alejandra Lizbeth Llontop Balarezo and Sharon Melissa Vilcarromero Ramírez, students of the Faculty of Law and Political Science of the National University of San Martín, are currently carrying out an investigation entitled: Protection and welfare of domestic animals with respect to the policies implemented by the Provincial Municipality of San Martín, Tarapoto, 2020.
Therefore, given his extensive knowledge and his excellent career as a professional in the field, currently occupying the post of Deputy Manager of Local Economic Development of the Provincial Municipality of San Martín, we would like to express our wish for him to participate in an interview which will be of great interest and support for the development of our research. This interview will allow us to delve into the knowledge necessary to obtain objective conclusions based on the experience of a professional with training in the field.
The interview lasts an average of one hour. Together with this interview, we provide you with our interview guide so that you can take into account the questions that will be addressed during the interview and we can benefit from your knowledge of the subject.
Thank you in advance.
Alejandra Lizbeth Llontop Balarezo Sharon Melissa Vilcarromero Ramírez

____
Alejandra Lizbeth Llontop Balarezo

____
Sharon Melissa Vilcarromero Ramírez

Annex 12: Invitation 02

Name of document: Law No. 30407: "Law on Animal Protection and Welfare".
09 September 2023
Tarapoto - Peru
Dear Elizabeth García Panduro, representative of the non-profit association Yupi Wasi:
First of all, a cordial greeting to you. Alejandra Lizbeth Llontop Balarezo and Sharon Melissa Vilcarromero Ramírez, students of the Faculty of Law and Political Science of the National University of San Martín, are currently carrying out an investigation entitled: Protection and welfare of domestic animals with respect to the policies implemented by the Provincial Municipality of San Martín, Tarapoto, 2020.
Therefore, as a representative of the non-profit association Yupi Wasi and your experience in the rescue and care of stray pets in the city of Tarapoto, we would like you to participate in an interview which will be of great interest and support for the development of our research. This interview will allow us to deepen the knowledge necessary to obtain objective conclusions based on your experience in the field of Animal Protection and Welfare.
The interview lasts on average one hour; together with this interview, we provide you with our interview guide so that you can take into account the questions that will be addressed during the interview and we can benefit from your knowledge of the subject.
Thank you in advance.

Alejandra Lizbeth Llontop Balarezo	Sharon Melissa Vilcarromero Ramírez

Alejandra Lizbeth Llontop Balarezo Sharon Melissa Vilcarromero Ramírez

Annex 13: Interview Guide

Good morning, Veterinarian Marco Antonio Sánchez Huaripata, we would like to thank you for the time you have given us to conduct this interview. We would also like to mention that the comments and information you provide us with will be very valuable for the thesis project to be carried out.

Objective: To analyse the information obtained from the interview with related professionals from the field of law, in order to question and determine the relationship between the "Protection and welfare of domestic animals with respect to the policies implemented by the Provincial Municipality of San Martín, Tarapoto, 2020", in order to carry out a research project.

Instructions: Read each question carefully and answer.

Information on the use and disclosure of data: I as the interviewee authorise this information to be disclosed in this research paper for academic purposes.

Duration of the interview: 1 hour.

Date: 05/07/2023

COORDINATOR PROFILE

Name: Marco Antonio Sánchez Huaripata

Profession: Veterinary Doctor

Position: DEPUTY MANAGER of Local Economic Development of the Provincial Municipality of San Martín.

Figure 1. Conducting the expert interview.

QUESTIONS
Academic Background
1.   What is your professional background?

My name is Marco Antonio Sánchez Huaripata, I am a veterinary doctor by profession, graduated from the National University of Cajamarca, I have a master's degree in sustainable animal production at the National Agrarian University of the Jungle of Tingo Maria, I have a master's degree in public health, I have doctoral studies in animal production, I am a teacher at the National University of San Martin - School of Veterinary Medicine, I am in charge of the Sub Management of Economic Development in the Provincial Municipality of San Martin and the position of sanitary inspector.

Environmental Law

2.   Could you comment on Law No. 30407 - Law on the Protection and Welfare of Animals, regarding domestic animals?

The law basically contemplates the conditions that a pet under our care must have, based on this, the law basically gives us that we must have the five freedoms as the main axis. We have to work on the basis of the five freedoms; the first one is freedom from hunger and thirst, to have adequate conditions, not to limit their functions as pets; then, we have a bad idea that we have humanised our pet and that is a bad practice that we suddenly have, but it has already been socialised, to celebrate our dog's birthday, to invite them to parades, to dress them and treat them. We have to take into account that it is a pet, we can suddenly consider it to a certain extent a member of the family, but also giving it its space as a pet, because by giving it its space as a pet we also give it its needs as a pet. Based on this, the law also contemplates animal comfort, the conditions they must have in their environment; by this I mean that if I have a flat I cannot have a Labrador, a Doberman, a German Shepherd; I must have the conditions, maybe a Chihuahua, a lap dog, whatever is best suited to our spaces. One also has to be aware of what I have and what I can provide or what I can give to my pet, because that is also part of animal welfare, the conditions I am going to provide.

3.   Has the Provincial Municipality of San Martin currently taken any action with respect to Law N°30407?

Of course, based on this framework law that we have, we have already enacted in 2020 the municipal ordinance 014-2020 which provides for responsible pet ownership here in San Martin. This ordinance regulates us on the ownership of our pets, in what conditions we have to have them, how we have to have them and the evaluations that we have to pass to have a pet; because, we have to have an authorization to have a pet and the authorization goes from the registration of our dog in the municipalities; because, the law also contemplates that the local governments are in charge of maintaining the pet registry and that also implies the type of pet that I am going to have, what does this mean, Our ordinance states that if I am going to have pets that are considered potentially dangerous, such as Argentine Dogo, Rottweilers, pit bulls and all their crossbreeds, according to the law, all the lines that I can obtain from them, I must also have special authorisations, including a psychological authorisation, to undergo a psychological evaluation. It must be taken into account that the pet is closely identified with a member of the family, mostly with the one who is in charge of all the care, but we have seen that there are people who misuse these pets, train them for improper purposes or sometimes keep them in unsuitable conditions; we must take into account that each species, each pet, has its own requirements, for example; I have a Weimaran, a Doberman, I will need a large space where it can run and develop; unlike a Chihuahua or a Maltese, the character of the animal depends a lot on the owner, but we must take into account that they also have an instinct for what they were created for. It says a lot about how we keep our pets, now we also have some inspections based on some photographic complaints that the neighbours make to us, we also go with the inspection area based on our ordinance to regulate this a little; now we no longer see in Tarapoto, at least, many dogs that live on the roof, but we do have dogs that still sleep at the door of the houses, on the public road and that is an issue that we also have to regulate in some way, obviously that is a very big public health problem, Because of the proliferation of certain pests that we have there, ticks and fleas, this leads to other zoonotic diseases such as rabies, we have to control all of this.

4.   Is there any decree or ordinance that the authorities of the Provincial Municipality of San Martin have issued regarding Animal Protection and Welfare? Please comment.

Of course, as I mentioned at the beginning, we have been working for some years on responsible pet ownership, and that is why Ordinance 014-2020-MPSM was born, which regulates responsible pet ownership, if we have it regulated here in the Province of San Martin, in Tarapoto we basically have an Ordinance that covers all of this.

5.   Is the Provincial Municipality of San Martin applying any policy in the district of Tarapoto in favour of the welfare and protection of domestic animals?

Effectively, we have been developing an inspection activity based on the ordinance that we have and based on the conditions that these pets have to have, previously, we have also had quite a few complaints about improper breeding of pets or animal abuse, which is also covered by the ordinance that we have been regulating continuously. We are constantly dealing with some complaints about dog bites as well, and that is where the ordinance comes in, we have to see what type of pet it has, if it is a crossbreed of a dog considered potentially dangerous. We are at a time of socialisation with the population so that they already have the corresponding permits to keep pets, what our inspection staff do is to see the conditions in which these pets are being kept, it is hard work, the city is large and sometimes it is a little difficult for the population to provide the conditions for pets, and that is what makes it difficult for us; Most of us have pets and not just one, some of us have two, three or even more, but this also implies a responsibility not only with the pet, it is also a social responsibility, because this is going to be reflected in the behaviour of our pets, what we do not want to see is dogs that are abandoned on public roads, or kittens, or other types of pets that due to lack of space or lack of conditions we abandon them, this is also what the ordinance regulates and this also gives us the right conditions to keep them.

6. What is the position of the Provincial Municipality of San Martin regarding the creation of an animal shelter?

We would like to have a municipal dog shelter or a municipal pet shelter, but at the moment circumstances still prevent us from doing so, basically due to logistical issues, but this has not limited us to the ideal; We are working, we have advanced some conversations with the animal protection associations, with them we are going to work together hand in hand to socialise this whole issue and start rescuing some dogs that are abandoned on public roads, we are not going to do it directly as a municipality, but through the associations and that will also be a great relief for some pets that live in conditions of neglect. The logistics issue involves a lot of economic issues and we would like to implement it, but we are going to have conversations later on with other institutions that have been working hand in hand with us to make this project viable and make it a reality; I believe that the Municipality and the University have a very good inter-institutional working agreement, we have been collaborating in many activities and we could talk to them to see how we can make this type of activity viable.

7. Is the Provincial Municipality of San Martin currently carrying out awareness campaigns?

Yes, based on the ordinance we have, we have already been carrying out some activities, from socialisation to responsible pet ownership; If you look at our Municipal website you will see that we are continually publishing some notices on responsible pet ownership, one of which is that if you take your pet out for a walk, don't forget to take it out with its muzzle, its leash and above all don't forget to take your waste bags with you, because you have to take into account that the same green spaces that our pets occupy are also occupied by our children and that is something that would play against the pets, but that is why the owners are urged to bring their bags so that they can pick up their pets' needs; We are constantly communicating on the Municipality's website through the social networks that we have.

8. What positive action is the Provincial Municipality of San Martin taking to help the situation of domestic animals?

We have been carrying out a series of activities in favour of our pets here in Sint Maarten, one of them is the registry, but as I mentioned, we are making a proper registry of our pets, both a national and an international registry, through a certification that will be granted to each of the pets, This means that for A or B I am going to take my dog on a trip, I have to obtain an authorisation to take it and the authorisation contemplates that I have to have a series of requirements, it is very easy, I have my dog or my pet registered in the Municipality, then, I simply enter the Municipality, I enter where my pet is and from there I download all the documents I need to be able to travel. The health control also implies that I will have to present my pet's health card, that it has its vaccinations up to date, its constant deworming, suddenly its medical check-up if necessary, and that will be included in the record that my pet will have at the Municipality. As I also mentioned, those who register their pet will be entitled to an international cynological registration certificate, we have a national one and we have an international one; The international one gives me the facility to take my dog abroad without any problem, it is as if I were going on a trip with just my ID card, because the registration not only contemplates a registry, but also having a history of our pet, not only through an ID card, but also by placing a microchip where all its data will be registered, that means that, anywhere in the world that I have a microchip reader, I will be able to identify my pet, we are talking about a microchip of data where the history will be registered, it is not a GPS. The history will include data such as: name of the owner, how to identify, how to locate him, in other words, how I will be able to provide all this data when I lose him, when I have to carry out any other procedure for my pet, I will have all the data at my fingertips. In reference to this project, we were talking about an urban nucleus considering Tarapoto, Morales and La Banda de Shilcayo, now we are talking from Juan Guerra to Cacatachi, but for the moment we have been working and we have had some meetings with the Municipality of Morales, which already has its Ordinance as well; With respect to responsible pet ownership and with La Banda de Shilcayo, I believe that their regulation is on its way, this makes it easier for us to work together and carry out much more efficient actions, not only with Morales and La Banda de Shilcayo, but we are also working together with OGESS - Bajo Mayo, which is directing a very good work plan, We are also working with the Professional Association of Veterinary Doctors in San Martin, where we have an agreement and where the first Municipal Veterinary Clinic is going to be built, for logistical reasons the Veterinary Medical Association is going to build and manage it, and we also have an agreement with the School of Veterinary Medicine - UNSM, what does this include?, Well, the school has already built a Veterinary Hospital, which means that everyone who needs a veterinary service will be able to have access to it at a totally social cost, since we are all suddenly not going to be able to take our pets to a clinic, but we will have other alternatives. We have to take into account that health costs are generally a little high, both for people and for pets, so we have seen the need to create this type of service through agreements and conversations that we have with the corresponding institutions; based on this, we also support the OGESS - Bajo Mayo, especially the canine rabies vaccination campaign, which is carried out every year here in the city of Tarapoto - San Martin. With regard to insurance and costs, we are not thinking of suddenly fragmenting or selecting pets, I believe that the social cost service will be available to any pet that wants it, whether or not they can pay for a private clinic, I will be able to access this service, I will simply take my pet and they will have the social costs both in the clinic and if necessary the clinic will refer them to the hospital, obviously all the social costs; Also, taking into account that the University also has an area of social projection, which is also what is going to make it easier for us to provide a service that is a little more accessible to the public, so that all pets can have access to a veterinary service or a consultation for a medical check-up if necessary.

Annex 14: Research instrument validation sheet by expert judgement

NATIONAL UNIVERSITY OF SAN MARTIN

FACULTY OF LAW AND POLITICAL SCIENCE

PROFESSIONAL SCHOOL OF LAW

Research: "Protection and welfare of domestic animals with respect to policies implemented by the
policies implemented by the Provincial Municipality of San Martin, Tarapoto, 2020".

1. GENERAL DATA.

1.1. Surname and first name of the expert: Jessica Elka Picón Pereda.

1.2. Academic degree of the expert: Lawyer.

1.3. Surname and first name of researcher : Alejandra Lizbeth Llontop Balarezo y
: Sharon Melissa Vilcarromero Ramírez

1.4. Name of instrument : Survey

1.5. Author of the instrument : Alejandra Lizbeth Llontop Balarezo and
: Sharon Melissa Vilcarromero Ramírez

II. CRITERIA FOR EVALUATING THE INSTRUMENT.

Indicators for assessing the instrument	Qualitative assessment	Deficient	Regular	Good	Very good	Excellent
	Quantitative assessment	1	2	3	4	5
Clarity	The items are formulated in appropriate language.					X
Objectivity	The instrument allows for the collection of observable data or behaviour.				X	
3. News	The instrument corresponds to the current state of knowledge.					X
4. Organisation	There is a logical organisation of the items.				X	
5.Sufficiency	The instrument assesses the quality and quantity dimensions of the variables.				X	
6. Intentionality	The instrument is adequate to achieve the objectives of the study.					X
7. Consistency	The instrument is based on scientific and theoretical aspects in line with the subject of the study.				X	
8. Coherence	The items relate to the variables dimensions and indicators.				X	X
9. Methodology	The instrument responds to the method, type, design and approach of the study.					X
10. Relevance	The items are suited to the type of research					X
TOTAL						

III. EXPERT OPINION

The instrument was approved. It was found to be valid and applicable for the type of research.

Tarapoto, 11 January 2023.

EXPERTO

Annex 15: Interview Guide

Good morning, Elizabeth García Panduro, representative of the non-profit association Yupi Wasi, we would like to thank you for the time you have given us to conduct this interview. We would also like to mention that the comments and information you provide us with will be very valuable for the thesis project to be carried out.

Objective: To analyse the information obtained from the interview with related professionals from the field of law, in order to question and determine the relationship between the "Protection and welfare of domestic animals with respect to the policies implemented by the Provincial Municipality of San Martín, Tarapoto, 2020", in order to carry out a research project.

Instructions: Read each question carefully and answer.

Information on the use and disclosure of data: I as the interviewee authorise this information to be disclosed in this research paper for academic purposes.

Duration of the interview: 1 hour.

Date: 09/09/2023

COORDINATOR PROFILE

Name: Elizabeth García Panduro

Profession: Clinical psychologist and integrative psychotherapist

Position: Representative of the non-profit association Yupi Wasi

Figure 1. *Conducting the expert interview.*
QUESTIONS

Academic Background

1.   What is your professional background?

Well, I am a clinical psychologist, I have a degree in clinical psychology, and I also trained in integrative psychotherapy, which is what I do most of my work, and I occasionally work in NGOs as a psychologist. In addition to that, I work with the Yupi Wasi association, which is very important to me, this issue of being a godmother and also a volunteer as an important part of human development of being responsible for pets, for animals, for me it is very important so, as my training is to work for people, to work with the fact of them as in their human quality, I think it is part of, it is very important.

Environmental Law

2.   Could you comment on Law No. 30407 - Law on the Protection and Welfare of Animals, regarding domestic animals?

Well, I know the law in name you could say, but I don't know what the application is, but I am glad to know that at least there is a law that is trying to promote the issue of protection. Although it is true that most people don't really know what it is about because they don't even know about it, at least there is, and I hope that little by little, through this type of project, through the associations, through the Municipality itself, more information about this law and above all how it is applied, because the law may be there but we often don't know how to proceed, what we have to do or what is considered an infraction or when I can make a complaint and that kind of thing. But at least it is there, the law is there, it is a first step.

3.   Has the Provincial Municipality of San Martin currently taken any action with regard to Law N°30407?

There is an ordinance, and I know that this new administration is a little more committed to this process. Although it is very slow due to the bureaucratic issue I imagine, but at least I know that it is working, at least for example in terms of inspection, I know that a person can go to the Municipality and report a case of abuse, abandonment, neglect or whatever it is that a person sees and considers that it is not right with respect to a pet, they can go, I don't know exactly which area, I don't know exactly what area, but I know that there is an area that oversees this, the Municipality sends them to check to see what is happening, if they are living in a bad situation, sometimes we know that they are living on the roof without water and without food, or also as we see them being raised in the street, I have also seen that some people let them loose to go for a walk and in some cases pets have been run over because of this; I know that a person can go and report it and there is a process, but I don't know what happens after that, I don't know what else happens after the report, if there is a fine, if there is a sanction of freedom, or if there is the fact of taking away the pet, I don't know more about that, but I do know that at least a prosecutor goes, gives certain, for example that you have to do this, this and this, your pet must improve in this, and then another possible visit, but from there I don't know what else. But I also know that the

Municipality is trying to work in other areas to improve in terms of responsible pet ownership and the punishment in case of negligence, abandonment or mistreatment.

4.   Is there any decree or ordinance that the authorities of the Provincial Municipality of San Martin have issued regarding Animal Protection and Welfare? Please comment.

I know, if I am not mistaken, correct me if I am wrong, the N°14-2020 that has to do a little, well I think it is more about dogs, I don't know if it is because it is the most common pet, but for example I know that it can be sanctioned the fact of not picking up the faeces, not walking without a leash or without a muzzle if it is necessary, I don't know how much people really know about this ordinance, beyond the number, which is important to know, but it is more important to know what it means, what this ordinance includes. I don't know if many people know about it, because there are many people who walk their dogs without a leash or, as I said, they just let them loose, which makes me laugh, but it really makes me wonder how they think of letting their dogs loose, because I don't even let a child loose when they go for a walk. So, I know that ordinance and I know more or less what it's trying to regulate, but with regard to dogs, I don't know if it's other types of animals or other types of pets. The most common ones are dogs and cats, but there are an infinite number of pets, the most common ones are dogs and cats, but even then I don't know about cats by far any kind of regulation.

5.   Is the Provincial Municipality of San Martin applying any policy in the district of Tarapoto in favour of the welfare and protection of domestic animals?

I have not heard anything about this, either through us or elsewhere, what I do know is that at the project level, is what I mentioned, the issue of seeking an agreement with the purpose of creating awareness in people to provide better quality of life and responsible pet ownership, I know that the Municipality is trying to put together perhaps part of a plan, but it is always very slow, it is creating agreements with the University, with the Veterinary College and I don't know if there are other private veterinaries that could be involved in order to create, for example, a clinic for basic care, obviously for dogs, for example sterilisation, vaccinations; sterilisation is not for stray dogs because sterilisation requires a post-operative period and if that animal is not going to have a place, then it is not done, that is why it is not done and if you take dogs from the street and sterilise them, no because they need a post-operative period; but at least there is a project for people of low income who cannot afford a vet but who want their dog to be looked after, they can have a place, they can have an option. I know that it is a project, I know that they are talking, trying to get more professionals involved, such as veterinary students, which I really think should have been a long time ago because we have the veterinary school, there are people who are studying veterinary medicine and can do this with the supervision of professionals obviously, but I do believe that it is viable and that this administration is at least moving it forward and that they are looking for this, I am also happy. I hope it will be soon and that it won't take so long, but that's what I know about it. I also know that they don't want to make a shelter because people don't have the awareness of responsibility and it would be like "there is a place where I can leave this dog, so I go and let it loose", that's one reason why not a shelter.

6.   What is the position of the Provincial Municipality of San Martin regarding the creation of an animal shelter?

About that, as I mentioned, I know that it is not in the plans of the Municipality because people, and I have seen it, people know that there is a shelter somewhere or it has happened with us, we are not a shelter, but they know that we rescue animals, all my pets that they have seen when they have arrived have been from the street, all of them, they see them as beautiful and everything, but when they arrived they were not like that; people see them and say "your dog is so beautiful, are you selling your dog? or "give me your dog", among other things, but they don't know that they haven't been like that and people often think that a shelter or the municipality should take care of them, but the reality is that there are animals in a bad state, with neglect or the fact of breeding them for sale, is precisely due to the irresponsibility of people, of human beings. So creating a shelter could unfortunately increase the likelihood of this happening more often. Here they have come to leave cats, when they see us passing in the street they ask us if we want another dog, as if we simply wanted to keep them, in other words there is a responsibility of care, feeding, their medication when they are sick, their vaccinations and all that, so I think that first we should work on the issue of awareness and responsibility, before thinking about a shelter, I think that the Municipality is very clear about that.

7.   Is the Provincial Municipality of San Martin currently carrying out awareness campaigns?

Not yet, and I have the hypothesis, let's say, that they do have the idea of doing it, but they are preparing to work on those issues that I mentioned about the agreements, so that the associations that work in rescue and all that, get involved in going and giving talks and raising awareness, but also the professionals are not involved in campaigns all of a sudden. I think that when there is a campaign, as there sometimes is, for vaccinations, cleaning, and that kind of thing that there is, many times I have taken them to my pets, there should also be a stand to touch on this issue, to raise awareness, why not see an animal as a gift that you are only going to have for a while, I sometimes hear comments like "I like them when they are puppies" but it is natural that they will grow, they are a living being, we cannot believe or think that they will always be a puppy, that they can always behave like a puppy, but they will grow. So sometimes for me it's like, what's in my head to think like that? "I would always like it to be a puppy" or until it is a puppy and then I put it away, I put it to one side, I throw it away, or if it is no longer useful for what I want, I leave it there. So I think that awareness and all that, we have to work on these issues as well, it is a responsibility of the time that it will last its life and that is less than ours, the fact of giving a chance to dogs that have a disability or that are already old, that are the most difficult to find an adopter, because they see a burden if they don't walk well, they see a burden if they are going to need treatment, but in reality to give them that quality of life I think it has to do with human quality as well. For example, my dogs, the "little one" as I call him, he has had physiotherapy, some people laughed because I took him to physiotherapy because he had an accident, but I think it is part of the fact that if your pet needs something and you can give it to him, and I think that is also important in terms of raising awareness, how many animals can I take care of? Because it does involve money, time, because you're not just going to keep them there, you have to take them for walks, you have to make them play because that's the way they are. So that's what it is, time, money and space, how many animals can you have. We would like to help more dogs but we can only do this, and this is part of education and being aware, and these are points that should also be touched on in the campaigns that are being carried out and in organisations such as schools, universities, like a talk about why this, why that, and why the other, because sometimes people get excited and want to have a dog, a cat, a rabbit, a hamster, but they don't know what their care involves, or they don't like the issue of bathing them, picking up their faeces, a lot of things that are involved, so we have to be aware that all these things are part of, and as far as I can.

8.   What positive action is the Provincial Municipality of San Martin taking to help the situation of domestic animals?

Well, I think that at least this issue of seeking to make agreements is a first step, and also the fact that I know that Rosa, who is the first to have started this association, has been called to talk, they have called her to tell her of their intentions, so I think that this is also a positive attitude, of wanting to be in contact with the entities or people who can help to promote all this animal protection and responsible ownership, So I see that at least the Municipality is looking to create spaces, perhaps it has many gaps, yes definitely, and I think it is more due to the bureaucracy that really is, I don't know how to say it, but at least they are making this movement that we hope will become more fluid and we can see more things, not only the associations or shelters, but also the population in general, know what to do and how to get involved, I hope it will be in the shortest possible time.

yes
I want morebooks!

Buy your books fast and straightforward online - at one of world's fastest growing online book stores! Environmentally sound due to Print-on-Demand technologies.

Buy your books online at
www.morebooks.shop

Kaufen Sie Ihre Bücher schnell und unkompliziert online – auf einer der am schnellsten wachsenden Buchhandelsplattformen weltweit! Dank Print-On-Demand umwelt- und ressourcenschonend produziert.

Bücher schneller online kaufen
www.morebooks.shop

info@omniscriptum.com
www.omniscriptum.com